Proteoforms - Concept and Applications in Medical Sciences

Edited by Xianquan Zhan

Published in London, United Kingdom

IntechOpen

Supporting open minds since 2005

Proteoforms - Concept and Applications in Medical Sciences
http://dx.doi.org/10.5772/intechopen.83687
Edited by Xianquan Zhan

Contributors
Fieke W. W Hoff, Anneke D. Dirkje van Dijk, Steven M. Kornblau, Xianquan Zhan, Shehua Qian, Olga Lima Tavares Machado, Jucelia Da Silva Araujo, Hartmut Schlüter, Siti Nurul Hidayah, Manasi Gaikwad, Laura Heikaus

Notice
Statements and opinions expressed in the chapters are these of the individual contributors and not necessarily those of the editors or publisher. No responsibility is accepted for the accuracy of information contained in the published chapters. The publisher assumes no responsibility for any damage or injury to persons or property arising out of the use of any materials, instructions, methods or ideas contained in the book.

First published in London, United Kingdom, 2020 by IntechOpen
IntechOpen is the global imprint of INTECHOPEN LIMITED, registered in England and Wales, registration number: 11086078, 7th floor, 10 Lower Thames Street, London, EC3R 6AF, United Kingdom
Printed in Croatia

British Library Cataloguing-in-Publication Data
A catalogue record for this book is available from the British Library

Additional hard and PDF copies can be obtained from orders@intechopen.com

Proteoforms - Concept and Applications in Medical Sciences
Edited by Xianquan Zhan
p. cm.
Print ISBN 978-1-83880-033-8
Online ISBN 978-1-83880-034-5
eBook (PDF) ISBN 978-1-83962-832-0

We are IntechOpen,
the world's leading publisher of
Open Access books
Built by scientists, for scientists

4,900+
Open access books available

124,000+
International authors and editors

140M+
Downloads

Our authors are among the

151
Countries delivered to

Top 1%
most cited scientists

12.2%
Contributors from top 500 universities

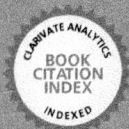

Interested in publishing with us?
Contact book.department@intechopen.com

Numbers displayed above are based on latest data collected.
For more information visit www.intechopen.com

Meet the editor

Xianquan Zhan received his MD and PhD in Preventive Medicine from West China University of Medical Sciences in 1989–1999. He received his post-doctoral training in Oncology and Cancer Proteomics at Central South University and University of Tennessee Health Science Center (UTHSC). He worked at UTHSC and Cleveland Clinic from 2001 to 2012, and achieved the rank of Associate Professor at UTHSC. Currently, he is a full professor at Central South University and Shandong First Medical University, and an advisor for MS/PhD graduate students and postdoctoral fellows. He is also a fellow of the Royal Society of Medicine, fellow of EPMA, a European EPMA National Representative, full member of the American Society of Clinical Oncology (ASCO), member of the American Association for the Advancement of Science (AAAS), editor-in-chief of *International Journal of Chronic Diseases & Therapy*, associate editors of *EPMA Journal* and *BMC Medical Genomics*, and guest editor of *Frontiers in Endocrinology* and *Mass Spectrometry Reviews*. Dr. Zhan has published more than 100 articles, seventeen book chapters, three books, and two US patents in the field of clinical proteomics and biomarkers.

Contents

Preface

The completion of the human genome sequence has driven the transition to the era of functional genomics, whose main contents are transcriptomics and proteomics. The human genome contains about 20,300 genes. However, the human transcriptome contains at least 100,000 transcripts, which is much more than the number of human genes due to RNA splicing and other factors during the transcription of a gene to RNA. Thus, multiple transcripts are often derived from one same gene. Each transcript guides the ribosome to synthesize an amino acid sequence of a protein. The synthesized protein in the ribosome must be translocated and redistributed to the appropriate locations to form a special conformation and interact with surrounding molecules, namely a complex, to exert its biological functions. Also, protein is modified by many post-translational modifications (PTMs) and even unknown factors in the process of translocation and redistribution. There are about 400–600 PTMs in the human body, which are the main factors for the complexity and diversity of proteins, namely, protein species or proteoforms. Thus, multiple proteoforms are often derived from one same transcript. About 1,000,000 proteoforms are estimated to exist in the human body. Actually, a protein is a set of proteoforms. The different proteoforms derived from one gene might have different conformations and functions. A proteoform is the final functional performer of a gene. Proteoforms are the basic units in a proteome. Each proteoform has its own copy number or abundance. Therefore, proteoforms further enrich the concept and content of a proteome. Studies on proteoforms will offer much more in-depth insights into a proteome, which will directly lead to the discovery of reliable biomarkers for accurate understanding of molecular mechanisms, the discovery of effective therapeutic targets, and for effective prediction, diagnosis, and prognostic assessment.

This book focuses on the concept of proteoforms, technologies to study proteoforms, and applications of proteoforms. Chapter 1 addresses the complete concept of proteoforms, and compares the methods of "top-down" mass spectrometry and two-dimensional gel electrophoresis-liquid chromatography-liquid chromatography-mass spectrometry (2DE-LC/MS). Chapter 2 examines the general concepts of proteoforms and the methodological process for identifying them. Chapter 3 describes in detail how to prepare proteoforms of therapeutic proteins for top-down mass spectrometry analysis, which opens up an area for the application of proteoforms in medical science. Chapter 4 discusses prolactin proteoform pattern alteration in human pituitary adenomas compared to control pituitary tissues, which provides an example of the application of proteoforms in clinical study. Chapter 5 covers proteoforms in acute leukemia, specifically evaluation of age- and disease-specific proteoform patterns.

This book presents new advances in the concept, methodology, and applications of proteoforms. However, this book contains only a fraction of the very important proteoform studies in medical sciences, which we hope will stimulate and encourage researchers who study proteoforms to come forward with scientific merits and clinical practice of proteoforms. We strongly believe that proteoform study will bring a brighter future for medical sciences and clinical practice.

Xianquan Zhan, MD, PhD.
Professor of Cancer Proteomics and PPPM,
University Creative Research Initiatives Center,
Shandong First Medical University,
Shandong, China

Chapter 1

Introductory Chapter: Proteoforms

Xianquan Zhan

1. Introduction

The completion of human genome sequence has driven the research focusing from structural genomics to functional genomics. Transcriptomics and proteomics are two main contents in the era of functional genomics. Human genome contains about 20,300 genes [1]. However, RNA splicing and other factors in the transcriptional process from a gene to RNA result in multiple transcripts that are derived from the same single-one gene. Thus, human transcriptome is estimated to contain at least 100,000 transcripts, much more than the number of human genes. Each transcript guides the ribosome to synthesize an amino acid sequence of a protein. The synthesized protein in the ribosome must be translocated and redistributed to the appropriate locations to form a special conformation and interact with surrounding molecules, namely, a complex, to exert its biological functions. Also, protein is modified by many posttranslational modifications (PTMs) and even unknown factors in the process of translocation and redistribution. An estimated 400–600 PTMs in human body are the main factors to cause the complexity and diversity of proteins, namely, protein species [2, 3] or proteoforms [4, 5]. Thus, multiple proteoforms are often derived from one same transcript, and it is estimated that human proteome contains at least 1,000,000 proteoforms [6]. A proteoform is the basic unit in a proteome, and it is defined as its amino acid sequence + PTMs + spatial conformation + localization + cofactors + binding partners + a function (**Figure 1**), which is the final functional performer of a gene [6]. A protein is an umbrella term for all proteoforms coded by the same gene. Moreover, the different proteoforms derived from one same gene might have different conformation and functions. Each proteoform has its own copy number or abundance, which can be quantified between given conditions [4]. Studies on proteoforms will offer much more in-depth insights into a proteome, which will directly lead to the discovery of reliable biomarkers to understand accurate molecular mechanisms, the discovery of effective therapeutic targets, and for effective prediction, diagnosis, and prognostic assessment.

It is a big challenge in the methodology to study the over millions of human proteoforms [1, 6]. The common bottom-up mass spectrometry (MS)-based strategies cannot identify proteoforms, which in fact only identify protein-coded genes, a protein group. This type of method includes stable isotope-labeled two-dimensional liquid chromatography-tandem mass spectrometry (2DLC-MS/MS) and stable isotope-free 2DLC-MS/MS, which only identify peptides and PTMs (**Figure 2**) [6]. Top-down MS-based strategies have been developed to identify proteoforms [7–9]. This type of method can identify proteoforms, which obtains the proteoform message including the amino acid sequence and PTMs. However, the obtained message of proteoform is only partial information of the above defined proteoform. Also, the protein must be purified prior to MS analysis, with different types of protein isolation techniques such as capillary zone electrophoresis (CZE) and liquid

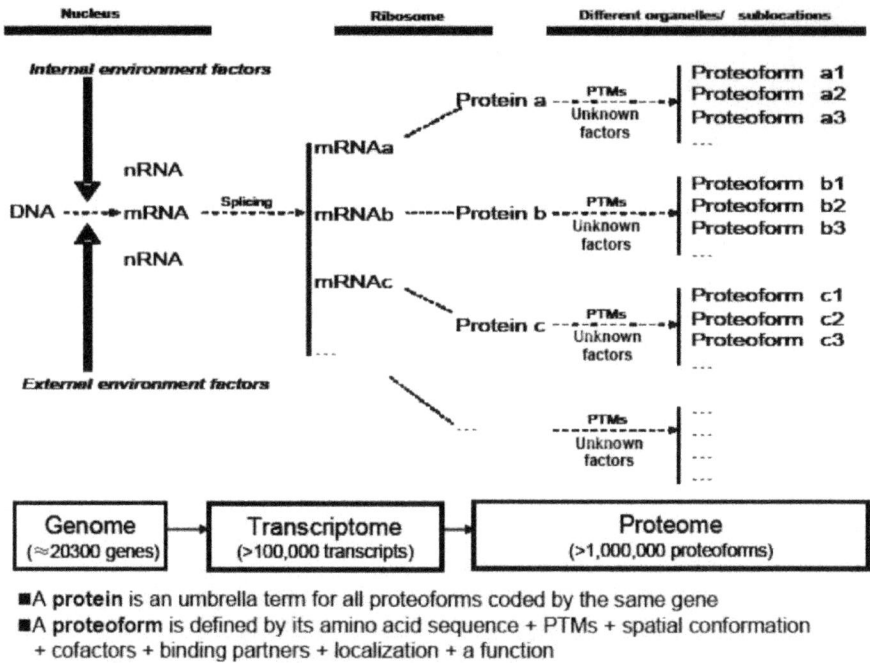

- A **protein** is an umbrella term for all proteoforms coded by the same gene
- A **proteoform** is defined by its amino acid sequence + PTMs + spatial conformation + cofactors + binding partners + localization + a function

Figure 1.
The concept and formation model of proteoform. (Reproduced from Zhan et al. [1, 6], copyright permission with open access policy.)

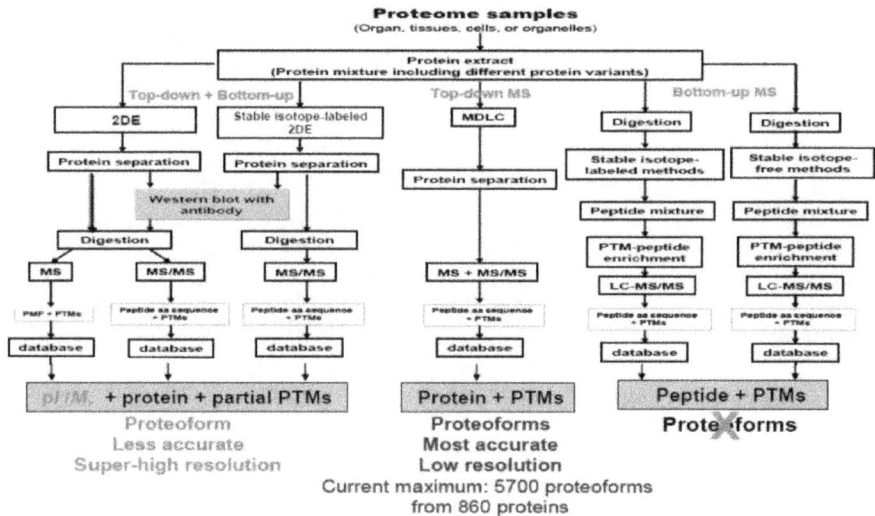

Figure 2.
The methods to study proteoforms. (Reproduced from Zhan et al. [1, 4, 6], copyright permission with open access policy.)

chromatography (LC) [10, 11]. Another drawback is the low ratio of signal to noise (S/N) in the MS analysis. All of those factors result in a relative low throughput in identification of human proteoforms. Currently the maximum throughput of top-down MS is up to 5700 proteoforms corresponding to 860 proteins (**Figure 2**) [6]. Two-dimensional gel electrophoresis (2DE)-liquid chromatography-MS

(2DE-LC/MS) strategy combines the top-down technique (2DE) and bottom-up technique (LC/MS), which is currently superhigh-throughput method to identify the large-scale proteoforms [1, 6, 12, 13]. With the innovating concept and practice of 2DE, 2DE is a real prefractionation method, which can effectively recognize isoelectric point (pI) and the relative mass (Mr)—two essential parameter of a proteoform; each 2D gel spot contains over 50 to several hundred proteoforms, and most of proteoforms are low-abundance. Currently, the largest 2D gel is 30 cm x 40 cm, which can separate 10,000 2D gel spots; thus at least 500,000 or 1,000,000 proteoforms can be identified. LC/MS can identify protein sequences and partial PTMs (**Figure 2**) [1, 13]. 2DE-LC/MS has great potential in analysis of large-scale proteoforms. 2DE-LC/MS and top-down MS are complementary in the achievement of maximum coverage of human proteoforms in a proteome.

Proteoform is the final functional format of a protein coded by a gene, which has important scientific merits in the fields of life sciences and medical sciences, and it is the research hot spot and international scientific frontiers. In the past 1–2 years, one has gradually paid more attention to the proteoform study. A total of 532 publications can be obtained through searching in the PubMed dataset with the keyword "proteoform or proteoforms." For example, 24 growth hormone (GH) proteoforms were identified with 2DE-LC/MS in human pituitary tissues [14], and 20 and 22 kDa GH proteoforms functioned in different signaling profiles. Six pro-lactin (PRL) proteoforms were identified with 2DE-LC/MS and 2DE-Western blot in human pituitary tissues, and the proportional ratio of six PRL proteoforms were significantly different among different subtype nonfunctional pituitary adenoma relative to control pituitary tissues [15]. The six PRL proteoforms bind to different long or short PRL receptors to exert their functions. A total of 3090 proteoforms were identified with liquid chromatography-MS (LC/MS), and 417 proteoforms were identified with sheathless CZE-MS, in seminal plasma [10]. A total of 3028 proteoforms corresponding to 387 proteins from *E. coli* cells were identified with coupling size exclusion chromatography (SEC) to CZE-activated ion electron transfer dissociation (CZE-AI-ETD) [16]. Human sperm protamine proteoforms were identified with a combination of top-down and bottom-up MS [17]. The glioblastoma [12, 13] and pituitary adenoma [13–15] tissue proteoforms were investigated with 2DE-LC-MS/MS. Proteoforms were identified from several cell lines (HepG2, glioblastoma, LEH) with 2DE-LC/MS [18]. Also, proteoform dynamics is also investigated underlying the senescence associated secretory phenotype [19].

In summary, development of proteoforms or protein species significantly enriches the concept of proteome, which is the next-generation research direction in the field of proteomics. 2DE-LC/MS and top-down MS are the complementary method to study the large-scale proteoforms. In-depth investigating proteoforms in a proteome with different pathophysiological conditions will directly cause to deeply understand disease molecular mechanisms, discover the reliable and effective therapeutic targets, and identify effective predictive, diagnostic, and prognostic biomarkers. Further, each proteoform is involved in a molecular network system and has multiple PTMs. It is the research hot spot how different PTMs competitively or synergistically affect proteoform structure and functions and their involved molecular network system [20–24]. Molecular network-based proteoform pattern biomarkers will have more important scientific merits.

Proteoforms are involved in the entire life science and medical sciences. This book contains only a fraction of the important frontier "proteoforms," which serve as a spur to stimulate and encourage researchers who study proteoforms to come forward with its scientific merits to research and clinical practice. This book will focus on the concept of proteoform, technologies to study proteoforms, and applications of proteoforms.

Acronyms and abbreviations

CZE	capillary zone electrophoresis
GH	growth hormone
LC	liquid chromatography
Mr.	relative mass
MS	mass spectrometry
MS/MS	tandem mass spectrometry
pI	isoelectric point
PRL	prolactin
PTM	posttranslational modification
S/N	ratio of signal to noise
2DE	two-dimensional gel electrophoresis
2DLC	two-dimensional liquid chromatography

Author details

Xianquan Zhan[1,2,3]

1 University Creative Research Initiatives Center, Shandong First Medical University, Shandong, China

2 Key Laboratory of Cancer Proteomics of Chinese Ministry of Health, Xiangya Hospital, Central South University, Changsha, China

3 State Local Joint Engineering Laboratory for Anticancer Drugs, Xiangya Hospital, Central South University, Changsha, China

*Address all correspondence to: yjzhan2011@gmail.com

IntechOpen

References

[1] Zhan X, Li N, Zhan X, Qian S. Revival of 2DE-LC/MS in proteomics and its potential for large-scale study of human proteoforms. Med One. 2018;**3**:e180008. DOI: 10.20900/mo.20180008

[2] Jungblut PR, Holzhütter HG, Apweiler R, Schlüter H. The speciation of the proteome. Chemistry Central Journal. 2008;**2**:16. DOI: 10.1186/1752-153X-2-16

[3] Schlüter H, Apweiler R, Holzhütter HG, Jungblut PR. Finding one's way in proteomics: A protein species nomenclature. Chemistry Central Journal. 2009;**3**:11. DOI: 10.1186/1752-153X-3-11

[4] Zhan X, Long Y, Lu M. Exploration of variations in proteome and metabolome for predictive diagnostics and personalized treatment algorithms: Innovative approach and examples for potential clinical application. Journal of Proteomics. 2018;**188**:30-40. DOI: 10.1016/j.jprot.2017.08.020

[5] Smith LM, Kelleher NL. Consortium for top down proteomics. Proteoform: A single term describing protein complexity. Nat. Methods. 2013;**10**(3):186-187. DOI: 10.1038/nmeth.2369

[6] Zhan X, Li B, Zhan X, Schlüter H, Jungblut PR, Coorssen JR. Innovating the concept and practice of two-dimensional gel electrophoresis in the analysis of proteomes at the proteoform level. Proteomes. 2019;7(4):36. DOI: 10.3390/proteomes7040036

[7] Chaffer LV, Millikin RJ, Miller RM, Anderson LC, Fellers RT, Ge Y, et al. Identification and quantification of proteoforms by mass spectrometry. Proteomics. 2019;**19**(10):e1800361. DOI: 10.1002/pmic.201800361

[8] Cupp-Sutton KA, Wu S. High-throughput quantitative top-down proteomics. Molecular Omics. 2020. DOI: 10.1039/c9mo00154a

[9] Shaw JB, Liu W, Vasil Ev YV, Bracken CC, Malhan N, Guthals A, et al. Direct determination of antibody chain pairing by top-down and middle-down mass spectrometry using electron capture dissociation and ultraviolet photodissociation. Analytical Chemistry. 2020;**92**(1):766-773. DOI: 10.1021/acs.analchem.9b03129

[10] Gomes FP, Diedrich JK, Saviola AJ, Memili E, Moura AA, Yates JR. EThcD and 213 nm UVPD for top-down analysis of bovine seminal plasma proteoforms on electrophoretic and chromatographic time frames. Analytical Chemistry. 2020;**92**(4):2979-2987. DOI: 10.1021/acs.analchem.9b03856

[11] Melby JA, Jin Y, Lin Z, Tucholski T, Wu Z, Gregorich ZR, et al. Top-down proteomics reveals myofilament proteoform heterogeneity among various rat skeletal muscle tissues. Journal of Proteome Research. 2020;**19**(1):446-454. DOI: 10.1021/acs.jproteome.9b00623

[12] Peng F, Li J, Guo T, Yang H, Li M, Sang S, et al. Nitroproteins in human astrocytomas discovered by gel electrophoresis and tandem mass spectrometry. Journal of the American Society for Mass Spectrometry. 2015;**26**(12):2062-2076. DOI: 10.1007/s13361-015-1270-3

[13] Zhan X, Yang H, Peng F, Li J, Mu Y, Long Y, et al. How many proteins can be identified in a 2-DE gel spot within an analysis of a complex human cancer tissue proteome? Electrophoresis. 2018;**39**:965-980. DOI: 10.1002/elps.201700330

[14] Zhan X, Giorgianni F, Desiderio DM. Proteomics analysis of growth hormone isoforms in the human pituitary. Proteomics. 2005;**5**(5): 1228-1241. DOI: 10.1002/pmic. 200400987

[15] Qian S, Yang Y, Li N, Cheng T, Wang X, Liu J, et al. Prolactin variants in human pituitaries and pituitary adenomas identified with two-dimensional gel electrophoresis and mass spectrometry. Frontiers in Endocrinology. 2018;**9**:468. DOI: 10.3389/fendo.2018.00468

[16] McCool EN, Lodge JM, Basharat AR, Liu X, Coon JJ, Sun L. Capillary zone electrophoresis-tandem mass spectrometry with activated ion electron transfer dissociation for large-scale top-down proteomics. Journal of the American Society for Mass Spectrometry. 2019;**30**(12):2470-2479. DOI: 10.1007/s13361-019-02206-6

[17] Soler-Ventura A, Gay M, Jodar M, Vilanova M, Castillo J, Arauz-Garofalo G, et al. Characterization of human sperm protamine proteoforms through a combination of top-down and bottom-up mass spectrometry approaches. Journal of Proteome Research. 2020;**19**(1):221-237. DOI: 10.1021/acs.jproteome.9b00499

[18] Naryzhny SN, Zorina ES, Kopylov AT, Zgoda VG, Kleyst OA, Archakov AI. Next steps on in silico 2DE analyses of chromosome 18 proteoforms. Journal of Proteome Research. 2018;**17**(12):4085-4096. DOI: 10.1021/acs.jproteome.8b00386

[19] Doubleday PF, Fornelli L, Kelleher NL. Elucidating proteoform dynamics underlying the senescence associated secretory phenotype. Journal of Proteome Research. 2020;**19**(2):938-948. DOI: 10.1021/acs. jproteome.9b00739

[20] Long Y, Lu M, Cheng T, Zhan X, Zhan X. Multiomics-based signaling pathway network alterations in human non-functional pituitary adenomas. Frontiers in Endocrinology. 2019;**10**:835. DOI: 10.3389/fendo.2019.00835

[21] Lu M, Zhan X. The crucial role of multiomic approach in cancer research and clinically relevant outcomes. The EPMA Journal. 2018;**9**(1):77-102. DOI: 10.1007/s13167-018-0128-8

[22] Zhan X, Long Y. Exploration of molecular network variations in different subtypes of human non-functional pituitary adenomas. Frontiers in Endocrinology. 2016;7:13. DOI: 10.3389/fendo.2016.00013

[23] Cheng T, Zhan X. Pattern recognition for predictive, preventive, and personalized medicine in cancer. The EPMA Journal. 2017;**8**(1):51-60. DOI: 10.1007/s13167-017-0083-9

[24] Zhan X, Desiderio DM. Editorial: Molecular network study of pituitary adenomas. Frontiers in Endocrinology. 2020;**11**:26. DOI: 10.3389/ fendo.2020.00026

Chapter 2

Proteoforms: General Concepts and Methodological Process for Identification

Jucélia da Silva Araújo and Olga Lima Tavares Machado

Abstract

The term proteoform is used to denote all the molecular forms in which the protein product of a single gene can be found. The most frequent processes that lead to transcript modification and the biological implications of these changes observed in the final protein product will be discussed. Proteoforms arising from genetic variations, alternatively spliced RNA transcripts and post-translational modifications will be commented. This chapter will present an evolution of the techniques used to identify the proteoforms and the importance of this identification for understanding of biological processes. This chapter highlights the fundamental concepts in the field of top-down mass spectrometry (TDMS), and provides numerous examples for the use of knowledge obtained from the identification of proteoforms. The identification of mutant proteins is one of the emerging areas of proteogenomics and has the potential to recognize novel disease biomarkers and may point to useful targets for identification of therapeutic approaches.

Keywords: post-translational modifications, top-down, mass spectrometry, proteomic experiments, clinical application of proteoform

1. Introduction

A surprise from the human genome project was the identification of 23,000 genes, far fewer than the estimated 100,000. Some events create distinct proteins that articulate various biological processes from cell signaling to genetic regulation. Thus, a single gene by allelic variations, alternative splicing and other pre-translational mechanisms, such as post-translational modifications (PTMs), conformational dynamics and functioning, may generate specific molecular forms of proteins, named "proteoforms," with different structures and different functions. Proteoforms or protein species as previously defined [1] could be identified by proteomics experiments, which include quantification of protein abundance, investigation of changes in protein expression, characterization of post-translational modifications (PTMs), identification of protein-protein interactions, a measure of isoform expression, turnover rate and subcellular localization [2]. Frequent modifications that produce proteoforms are presented in **Figure 1.**

IntechOpen

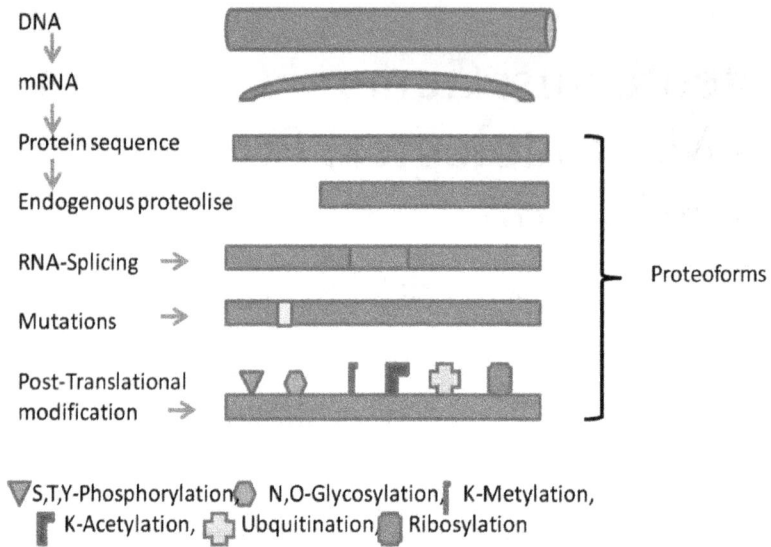

Figure 1.
Types of proteoforms: RNA splicing and mutations.

2. Proteomic experiments

The advance of genomics enabled the sequencing of the genes of an organism, but this does not inform which proteins may be present or how they are modified in specific situations. The proteomics analyses begin with the combination of multi-dimensional separation included the chromatographic and gel electrophoresis techniques and the ability of mass spectrometry to identify and to precisely quantify the proteins. Many different technologies have been and are still being developed to get the information contained in proteoforms. The high-precision mass spectrometric measurements as the tandem mass spectrometry (MS/MS or MS2 (peptide mass fingerprinting)) can provide structural information on molecular ions that can be isolated and fragmented [3, 4]. Mass spectrometry-based proteomics can be carried out in a bottom-up or top-down approach.

2.1 Bottom-up proteomics

The bottom-up proteomics also termed "shotgun proteomics," when the bottom-up analysis is performed on a mixture of proteins, has traditionally been used. In this approach, proteins that could be a simple or complex mixture are digested by chemical or enzymatic digestion to generate peptides that are analyzed by way of MS2. It is generally applied to identify and characterize many peptides in a mixture and deduce the identity of the protein to exist in the sample [5, 6]. In strategy bottom-up, the peptides mixture resulting of the digestion is fractionated and subjected to multidimensional liquid chromatography, which consists of prefractionating of peptides first according to their net charge using strong cation exchange chromatography and second, according to their hydrophobicity by reversed phase liquid chromatography (RP-LC) coupled online with a mass spectrometer [7]. The peptides fragmented within the mass spectrometer will provide product-ion mass spectra which are compared with in silico-generated MS/MS of the same mass encoded in a protein database. Proteins present in the sample are then inferred from the identified peptides [8].

This approach brings up several disadvantages: the protein inference process can be complicated because proteins often contain homologous sequence regions and the peptides cannot be either uniquely assigned to a single protein, the same peptide might have originated from multiple protein isoforms and/or from distinct functional pools of the same protein [2, 9]. The digestion of proteins can cause loss sequence variations or information regarding the original amino acid sequence and loss information relationship between the amino acid sequence and the PTMs belonging to specific proteoforms; thus, it is not capable of identifying proteoforms [6, 10]. Introducing the intact protein into the mass spectrometer eliminates these problems, the strategy used by the top-down mass spectrometry.

2.2 Top-down proteomics

Differently, of the bottom-up proteomics, the "top-down" approach involves direct separation and MS analysis of intact proteins, without previous proteolytic digestion. By this method, proteoforms can be characterized since the relationship between the amino acid sequence and the PTMs is preserved, and thus, such characterization provides a proteoform-specific understanding of biological phenomena [10]. In top-down proteomic, a specific proteoform of interest can be directly isolated and, subsequently, fragmented in the mass spectrometer by MS/MS strategies to map both amino acid variations to obtain information on protein masses [11]. The masses describe the complete amino acid sequence, including all post-translational modifications, structures, for successful identification [12, 13].

In this proteomic analysis, proteoforms are identified using precursor mass and fragmentation data. A precursor mass spectrum (MS1) of intact proteins is recorded; the most intense peaks are selected for fragmentation; and mass spectra (MS2) of the resulting fragment ions are acquired. On this account, both its intact and fragment ions' masses are measured, and the precursor masses and their isotopic distributions present a complex but detailed set of information. Upon fragmentation, terminal fragments represent potential cleavage site(s) or truncated proteoforms, while (internal) fragment ions can indicate modifications and, depending on the achieved sequence coverage, possible location. This approach routinely allows for 100% sequence coverage and full characterization of proteoforms [6, 13]. Top-down mass spectrometry has become the approach handy for the analysis of single proteins or simple mixtures of significant biological interest. Complexes proteomic samples require that they are fractionated before to introduction to the mass spectrometer. Many separation strategies can be applied before mass spectrometer only the last step, usually, the RPC coupled to mass spectrometry [6, 14].

Proteomic top-down may be denaturing or native. Denaturing top-down proteomics (dTDP), the procedure denaturing provides a powerful technique for characterizing individual proteins <30 kDa. In these studies, the proteins are denatured prior to their introduction into the mass spectrometer [15]. In this approach, protein interactions and quaternary conformations are disturbed by means of substance such as organic solvents, reducing agents, strong detergents, non-physiological pH, and/or physical method as heat and pressure [16]. TDP is the most disseminated and the scoring system used for identification and characterization of proteoforms. It has naturally been developed and tested using datasets derived from denatured top-down mass spectrometry experiments [17].

Native top-down proteomics (nTDP) has been used to characterize intact, non-covalently bound protein complexes biologically relevant as non-covalent protein-protein and protein-ligand interactions, providing stoichiometry and structural information since tertiary and quaternary structures of proteins are

Bottom-up proteomics

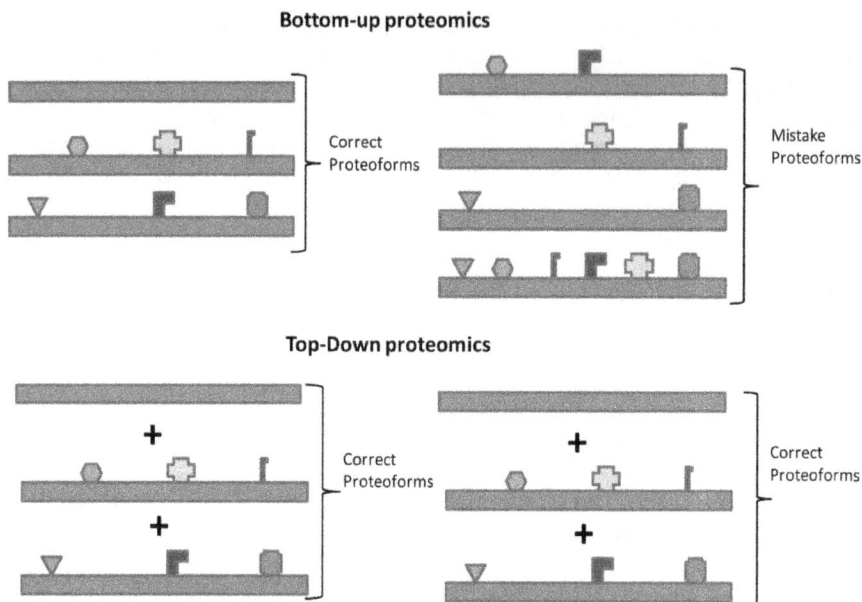

Top-Down proteomics

Figure 2.
Comparative scheme between top-down and bottom-up proteomics, showing the best indication of top-down use for the identification of proteoforms in complex protein mixtures.

maintained [6, 18]. The technique utilizes non-denaturing and non-reducing-buffer conditions during the electrospray ionization process which helps preserve the primary and quaternary compositions of proteins and their complexes for MS [19, 20]. In native proteoform approach, analytical platforms for high-resolution and liquid-phase separation of protein complexes are required prior to native mass spectrometry (MS) and MS/MS [21]. During, Escherichia coli proteome analysis, 144 proteins, 672 proteoforms, and 23 protein complexes were identified, coupling the size-exclusion chromatography and capillary zone electrophoresis-MS/MS [21]. Other separation techniques have been combined; coupled off-line ion-exchange chromatography or gel-eluted liquid fraction entrapment electrophoresis (GELFrEE), which demonstrates the compatibility of native GELFrEE with native and tandem mass spectrometry [21, 22]. An illustrative scheme of the elucidation of the proteoform structures by the bottom-up and top-down techniques is presented in **Figure 2**.

3. Separation techniques applied to proteoforms

The number of proteoform species in a proteome could be vast. Separating proteoforms is essential because many high-resolution mass spectrometers due to limited charge capacity have a finite ability to detect proteoforms. High-resolution separation techniques for complex protein samples are significant challenges of top-down proteomics. To optimize proteome coverage, separation, and multidimensional combinations, strategies are employed, thus, to reduce the complexity of the samples [23]. The strategies of separation of proteoforms are based on your intrinsic characteristics and physicochemical properties, such as mass/size, isoelectric point and hydrophobicity. Advances in instrumentation, chromatographic and electrophoretic separation strategies have been developed to separate intact proteins [20] since polyacrylamide two-dimensional gel electrophoresis (2D-PAGE) [24, 25] to the development of gel-eluted liquid fraction entrapment electrophoresis (GELFrEE), capillary zone electrophoresis (CZE) [20, 24–26]. Specific columns are also developed for classical methods of separation such as hydrophobic interaction chromatography (HIC), hydrophilic interaction chromatography (HILIC), reversed phase liquid chromatography (RPLC), chromatographic ion exchange (IEX) and size exclusion (SEC) [21–23, 27]. Some separations can be on-line with a mass spectrometer as separations chromatography and capillary electrophoresis, but many others can be applied off-line only [26]. Off-line separations approach, independent of the mass spectrometer, is flexible and diversified, allowing the use of diverse techniques of separations, although it is more laborious considering the time of collection and treatment of the fractions. Off-line separations system consists of three steps: separation of the sample compounds in the first dimension; a collection of different fractions for subsequent sample treatment; and injection of each of the fractions in the second dimension to be subject to analysis [20]. On-line separations, coupled directly to mass spectrometry, allow increased throughput and substantially reduce sample handling, but have limitations to sample loading, data acquisition and separation conditions [6, 26]. Many techniques of fractionation and separation can be combined to reduce the complexity of samples, and an off-line approach coupled with an on-line separation may be necessary since most proteomic samples have such complexity that they need multiple separation steps combined in multidimensional separations [27].

3.1 Sample preparation for top-down proteomics

The preparation of samples is one of the most critical steps for top-down proteomics. Conventional buffers for protein extracting as the detergents sodium dodecyl sulfate (SDS) and Triton X-100 are not compatible with MS [12]. Many methods for lysing samples use saline buffers, reducing agents, and protease and phosphatase inhibitors to extract proteoform avoiding alteration or degradation. Post extraction is needed to remove or replace nonvolatile salts that suppress MS signal by forming adducts to protein ions and increase the chemical noise [26]. The most strategies of proteins solubility for proteoforms studies, despite preserving many covalent interactions, denature proteins prior to MS and destroy important interactions such as protein-protein. In general, these interactions are essential for many different cellular processes [28]. In procedures for top-down native, the pH must be kept neutral and isolating, and fractionating of proteoforms complexes cannot contain denaturing agents such as strong detergents, reducing agents and organic solvents [29]. For these proteins to remain in their native states, a buffer is generally used for maintaining physiological ionic strength and neutral pH of the sample. To minimize noise associated with common buffer and others numerous interfering components, top-down sample cleanup methods should be applied; for example, protein precipitation and molecular weight cut-off ultrafiltration. Donnelly's et al. guide is one of the best-practice protocols for MS of intact proteins from mixtures of varying complexity [30].

3.2 Two-dimensional gel electrophoresis (2DGE)

2D-PAGE is an electrophoretic separation technique still used to separate intact proteins; the protein separation in 2D-PAGE is based on the isoelectric point and molecular weight (MW) of the proteins. This technique was introduced by O'Farrell [31], and it separates cellular proteins under denaturing conditions and enables the resolution of hundreds of proteins. In the first dimension, the separation is based on the proteoforms net electric charge (isoelectric point) of each protein and in the second dimension, in the presence of sodium dodecyl sulfate (SDS), proteins will be separated according to their molecular mass [31, 32]. The denaturing conditions introduced to O'Farrell [31] for first dimension comprise conducting a sample preparation, using high concentration molar of urea (9 mol/L), nonionic detergent (Nonidet NP-40) and a thiol reagent (2-mercaptoethanol), obtaining in this way an efficient separation of the proteins contained in the complex sample [33]. The use, in the first dimension, of tube gels and ampholytes to establish the pH gradient was replaced by the introduction of immobilized pH gradients (IPG strips). A significant advance on 2D-PAGE occurred with the development of the IPGs by [34] available in various ranges of pH and size. The IPG in polyacrylamide gels allows an efficient and reproducible separation of the proteins. In IPGs, the carrier ampholytes are attached to acrylamide molecules and cast into the gels to form a fixed pH gradient and covalently bound to a film backing. In this case, the buffering groups are grafted to the acrylamide gel matrix, the gradients cannot drift, and the gel slabs can be cut to narrow, usually 3 mm wide. Using IPG strips, the first-dimension separations are more reproducible and have high throughput and high resolution. The IPGs are much easier to handle, and there is the convenience provided by commercial production of IPG strips made [33, 35, 36].

3.3 Differential in gel electrophoresis (DIGE)

Conventional 2D gels were revolutionized with the introduction of the differential gel electrophoresis (DIGE), which allow the accurate and reproducible quantification of multiple samples by the relative intensity of fluorescent-dyed protein spots that are quantified within the same gel. Difference gel electrophoresis enables the accurate quantification of changes in the proteome, including proteoforms [37]. It is a strategy that has been developed for the quantitative analysis of intact proteins, and provides important information about changes caused by events such as truncation, degradation, genetic code variation, alternative splicing, post-translational processing and PTMs [37, 38]. The proteins in each sample are covalently tagged with different color fluorescent dyes, known as CyDye DIGE fluorescent. Fluorescent labeling of proteins is performed prior to 2D-DIGE, and then minimal labeling is often performed, such that <5% of proteins are labeled, thus reducing interference with downstream mass spectrometric analysis [39]. 2D DIGE involves the use of a reference sample, known as an internal standard, which comprises equal amounts of all biological samples in the experiment [39]. The major advantages of DIGE are the high sensitivity and linearity of the dyes utilized, its straightforward protocol, as well as its significant reduction of inter-gel variability, which increase the possibility to unambiguously identify biological variability and reduce bias from experimental variation [39, 40].

3.4 Gel-eluted liquid fraction entrapment electrophoresis (GELFrEE)

GELFrEE is a type of approach based on protein array developed to overcome the difficulties related to gel-based. It is one robust strategy that promotes size-based separation of proteoforms (applied to proteins 10–250 kDa) in the liquid phase with high resolution. GELFrEE is a electrophoresis to accommodate broad mass range separation of proteins, and the separation can be performed under denaturing or also been adapted for native-state size separations, where the tertiary and quaternary structures of the proteins are maintained [16, 22, 41]. In GELFrEE, the gel column is used to achieve electrophoretic separation of proteins, analogous to SDS-PAGE. The proteins are loaded onto the top of a tube containing poly-acrylamide gel; for separation to occur, a voltage is applied between the anode and cathode reservoirs which are then eluted into the liquid-phase for manual collection, securing that higher molecular weight proteins are not continually diluted and dispersed across many fractions [6]. The detergents incompatible with MS can be removed using organic solvent precipitation before online LC-MS. Spin columns are coupled in-line matrix removal platform to enable the direct analysis of samples containing SDS and salts detergents used in native mode [6, 42, 43]. This technique has the advantage of separating proteoforms over a wide mass range in short time and at high load, but there is the disadvantage of loss of resolution in the detergent removal stage; an acid labile surfactant may be an alternative to SDS [44]. Many combinations on-line or off-line GELFrEE with other fractionation techniques have been applied for optimal workflows for large-scale intact protein analysis. The fractions obtained from the electrophoretic step GELFrEE (for molecular-weight-based fractionation) are submitted to a second separation dimension. Li et al. [18] identified 30 proteins in the mass range of 30–80 kDa from *Pseudomonas aeruginosa*, fractionated by GELFrEE, analyzed by CZE-ESI-MS platform. However, the workflow of additional separation procedure most commonly performed is using a GELFrEE-LC-MS/MS [45–47].

3.5 Capillary zone electrophoresis

CZE has been the most common CE mode applied to the mass spectrometry of intact proteins. It is a method of proteoforms separation based on electrophoretic mobility differences that do not require a stationary phase [20, 48]. This approach provides fast and efficient separations. The sample is injected into an electroosmotic flow generated by the potential difference between two ends of a capillary filled with an aqueous solution, and the molecules are separated by the electrophoretic mobility difference. Capillary zone electrophoresis (CZE) offers alternative and high-capacity separation of proteoforms based on their sizes and charges, and are useful for the separation of high-mass proteoforms by not having stationary phase [49, 50]. CZE has been an alternative to RPLC; for example, CZE-MS interfaces have better sensitivity to detection, and it can produce more protein identifications from complex proteome samples than typical RPLC-MS [51]. The combination of methods CZE has led to efficient separation and highly sensitive detection of intact proteoforms with the benefit of low sample amount needed, inclusive for native proteomics, but some challenges still need to be overcome [20, 52].

3.6 Liquid chromatography systems

Liquid chromatography is the main proteomic approach used for protein separation in the mono- or multidimensional modes, which is ideally suited for proteomics because it can be interfaced with MS. The basic principle of chromatographic separation is the different affinity of analytes for the stationary and mobile phase. The LC-based separation methods have the advantage that they can be coupled directly with MS [53, 54]. Various orthogonal separation techniques using different stationary phase with different types interactions selectivity, two-dimensional LC separation (2D LC), and multidimensional LC separation (MDLC) are often combined to improve intact protein separation and proteoform coverage and to increase the dynamic range of detection. Multiple orthogonal separations include reversed phase (RP), ion exchange (IEX), size-exclusion (SEC), hydrophilic interaction liquid chromatography (HILIC) and hydrophobic interaction liquid chromatography (HIC).

3.6.1 Reversed phase liquid chromatography separation (RPLC)

The separation of the proteoforms in RPLC is based on their hydrophobicity using a non-polar stationary phase and a polar mobile phase; the analytes are subsequently eluted using increasing concentrations of organic solvents. The RPLC approach is widely used for complex intact protein sample separation and fractionation, and when coupled online with MS, it is the most prevalent approach for studying complex intact protein samples in top-down proteomics [13, 55]. Efficient separations to improve peak capacity have been achieved with the use of longer columns' smaller particle sizes in ultra-high pressure LC systems such as long column ultrahigh-pressure liquid chromatography (UPLC). Particle generally, either silica-bonded or polymeric-bonded octadecyl (C18), octyl (C8), or other shorter alkyl chains stationary phases are used such as C4 and C5 for intact protein separation [56–58]. The separation of the proteoforms in RPLC is based on their hydrophobicity using a non-polar stationary phase and a polar mobile phase; the analytes are subsequently eluted using increasing concentrations of organic solvents. Effective separations to improve peak capacity have been achieved with the use of longer columns smaller particle sizes in ultra-high pressure LC systems such as long column ultrahigh-pressure liquid chromatography (UPLC). Particle generally, either silica-bonded or polymeric-bonded octadecyl (C18), octyl (C8), or other shorter alkyl chains stationary phases are used such as C4 and C5 for intact protein separation [55–57]. Due to extreme complexities, limited sample loading amounts, and large dynamic ranges of intact protein samples, RPLC alone may not provide sufficient proteome coverage for top-down proteomics. One common way to increase peak capacity in RPLC and increase the proteome coverage is to include 2D RPLC or multiple orthogonal separation steps during analysis. Some high-resolution techniques combined with RPLC, for separation of proteins and proteoforms, for example, IEX-HILC-RPC/MS, high-pH and low-pH RPLC 2D (2D pH-RPLC-RPLC), are used for mass spectrometry compatible [58, 59].

3.6.2 Ion exchange chromatography (IEX)

Ion-exchange chromatography is a LC technique for proteins separation for top-down proteomics based on differences in charge of the analyte. IEX can be applied in cation- and anion-exchange modes. Increasing the ionic strength of the mobile phase is used to elute analytes from the charged stationary phase. The efficiency in the separation of proteins in IEX is related to conditions to salt concentrations and pH elution process applications that can be well versatile. This approach is often employed to carry out the first dimension followed by RP chromatography in the second dimension to 2DLC, or 3DLC strategy using, for example, IEX-HILIC-RPC/MS.

3.6.3 Size-exclusion chromatography (SEC)

Size-exclusion chromatography is part of the intact protein analysis workflow. The fractionation occurs in the difference in the accessibility of proteins to the intraparticle pore volume of the resin, in the non-adsorptive mode of solute interactions with the stationary-phase surface. The proteins migrate through a porous polymeric column and are separated by their hydrodynamic volume, with more abundant proteins eluting before smaller ones due to their lower accessibility to the interior of the packing materials. The selectivity is provided by the column, defined by the size of the intraparticle pore diameter; thus, the efficiency in the

SEC separation is mainly governed by the particle diameter [60, 61]. SEC has the advantage that can be realized in several types of solutions; however, it is not a high-resolution separation method, in addition to promoting the dilution of the sample. To increase the performance of SEC, different aspects with respect to column technology and instrumentation have been addressed. Huang et al. [62] developed a simple and efficient method SEC-based separation of proteins using RP columns (RP-based SEC performed). They have applied high concentrations of acetonitrile with trifluoroacetic acid as an acid modifier which prevented interactions between proteins and the stationary phase and allowed the RP column to act as an SEC column to separate proteins based on their molecular weight. This innovation showed that the RP-based SEC performed better than conventional SEC. Cai et al. [63] innovated the SEC-based separation. They developed a serial size exclusion chromatography (sSEC) strategy to enable high-resolution size-based fractionation of intact proteins. They combined SEC with different pore sizes in series and an increase in sufficient separation length, providing an extension of fractionation range and higher-resolution separation of proteins pool. This strategy of sSEC coupled to RPLC quadrupole-time-of-flight mass spectrometry provided improved proteome coverage [63].

3.6.4 Hydrophobic interaction chromatography (HIC)

HIC is a technique that separates proteins based on hydrophobicity with high resolution for the separation of intact proteins, main native conditions, and is an alternative MS-compatible LC if appropriate salt is used in the mobile phase [14, 59]. In this approach, protein's tertiary structure binds to a hydrophobic surface material in the presence of salt and then elutes in order of increasing surface hydrophobicity. The stationary phases used for HIC generally feature low density and moderate hydrophobic ligands, and resins that are less hydrophobic as compared to their counterparts used in RPLC, being the most.

3.6.5 Hydrophilic interaction chromatography (HILIC)

Hydrophilic interaction chromatography is a technique successfully applied to the separation of proteins and proteoforms. HILIC has the ability to retain and resolve highly polar compounds, based on a complex retention mechanism, involving hydrophilic partitioning and polar interactions; in other words, the analytes are eluted based on their hydrophilicity [64, 65]. In HILIC, the stationary phase is polar and often consists of a silica support that can be unmodified or modified with a polar surface chemistry, such as zwitterionic sulfoalkylbetaine, amide, diol, and aminopropyl; and the mobile phase consists of water and 60–95% of an aprotic, miscible organic solvent, usually acetonitrile (ACN) or acetone, with at least 3% of water. An organic solvent is used in loading HILIC columns to drive hydrophilic portions of proteins to interact with a hydrophilic stationary phase. Elution using a gradient from an organic solvent to an aqueous buffer allows desorption and elution of proteins from the column [66–68]. HILIC is MS-compatible LC technique for protein analysis. Therefore, coupling HILIC techniques in online or off-line two-dimensional LC workflows has increased the efficiency on the LC-MS analysis of complex protein samples, HILIC to be complementary and orthogonal to RPLC [65, 69]. Gargano et al. [70] implemented a capillary HILIC-MS method that can be used as a high-resolution approach to separate complex mixtures of proteins using wide mobile-phase gradients. Salt-free pH-gradient IEX-HILIC was used as the second dimension for separating differentially acetylated/methylated intact protein isoforms in histone family and combined this separation with RPLC online in the first dimension to better separation and characterization of intact histones [71].

4. Mass spectrometry

Proteomic experiments, MS based on comprehensive and total characterization of proteoform from a biological system, besides efficient separation, employ a combination of sensitive detection and accuracy of intact proteins. The technology for identification by MS to top-down proteomics has gained impulse. The accuracy of mass spectrometric characterization of polypeptides involves improvement on ionization, fragmentation and detection conditions. Tandem MS can confirm the protein identification based on the daughter ions and characteristics of the obtained peptide map and primary structure, which thereafter provide exact localization of post-translational or other modification sites. Data-independent acquisition (DIA) methods have been alternatively used to analyze proteoforms particularly suited to the study of PTMs [72]. DIA focuses on the identification and quantitation of fragment ions that are generated from multiple peptides contained in the same selection window of several to tens of m/z, that is, the fragmentation spectra of all the peptides are acquired in each cycle time without any preselection of the precursor ions [73].

The mass spectrometers are compounded basically into a sample inlet, an ion source, a mass analyzer and a detector [74, 75]. Although MS appeared more than a century ago, its application to protein analysis began in the 1990s, because existing ion sources only allowed the ionization and analysis of inorganic molecules and small organic molecules and proteins are not easily transferred to the gas phase and ionized by the size [76, 77]. Advancement of mass spectrometry technology occurred with the new instrumentation ionizer, matrix-assisted laser desorption/ ionization (MALDI) and electrospray ionization (ESI) [78–80]. The development of the mass analyzer applied to analyze intact proteins contributed to the mass spectrometry identification of the proteoforms. Mass analyzers with a high level of resolving power and sensitivity as time-of-flight (TOF), Orbitrap, Fourier Transform Ion Cyclotron Resonance (FT-ICR), or the combination of multiple mass analyzers in series, created a powerful tool for top-down MS characterization of proteoforms [78, 81, 82]. Most top-down proteomics (TDP) studies have used some form of tandem-MS fragmentation techniques, for intact proteins sequencing with greatly resolving power and high mass accuracy as: collisionally activated dissociation (CAD), collision-induced dissociation (CID), electron transfer dissociation (ETD), electron-capture dissociation (ECD), higher-energy collisional dissociation (HCD), infrared multiphoton dissociation (IRMPD) and ultraviolet photodissociation (UVPD). These examples of fragmentation strategies can provide additional information on the amino acid sequence and PTMs for identification of proteoforms [74, 75, 79, 83]. The mass spectrometer sample introduction can be through the traditional RPLC-MS, by CZE-MS or embedded in a matrix on a target plate [74, 84]. Mass spectrometers that use different types of analyzers for the first and second stages of mass analysis (hybrid MS instruments) are employed to maximize proteoform characterization top-down MS-based. Still, software tools for the identification and quantification of proteoforms need to be continuously developed to keep up with a demand to quickly and automatically analyze the data generated. Many a comprehensive proteoform software tools for proteoform identification and construction of proteoform families are freely available: MASH Suite, MetaMorpheus, MSPathFinder, Proteoform Suite, TDPortal, TopMG and TopPIC

[13, 20] that can be implemented into current top-down workflows consecutive at complete and accurate databases.

A common material used is either surface-modified silica or polymeric particles coated with short aliphatic groups n-alkyls (propyl, butyl, hexyl, or octyl chains), phenyl and others [61, 85]. HIC separation methods have been evaluated and optimized as complementary selectivity to RPLC, which offer efficient separation for highly orthogonal HIC-RPLC for top-down proteomics [14, 27].

5. Clinical applications for proteoform identification

Several studies are carried out aiming to find markers for pathophysiology process of Alzheimer's disease (AD), cancer [86], type 2 diabetes, and chronic alcohol abuse, among other diseases. The identification of proteoforms associated with different diseases will undoubtedly be an essential dividing mark for early diagnosis, prevention and treatment. Some examples for proteoform identification applications as apolipoproteins proteoforms, B-type natriuretic peptide (BNP), disorders of glycosylation, detection of structural changes in transthyretin, hemoglobin proteoforms, cystatin C-truncated proteoforms, C-reactive protein, vitamin D-binding protein, transferrin and immunoglobulin G (NISTmAb) were discussed [86, 87]. In the last 30 years, since the MALDI and ESI approaches were developed, only about a dozen of mass spectrometry protein identification tests have been described. Here, we present studies involving Alzheimer's disease and alterations in the levels of apolipoproteins associated with lipid metabolism.

5.1 Alzheimer's disease

In the diagnosis of Alzheimer's disease (AD), quantification of total Tau protein (T-tau), threonine-phosphorylated Tau181 form (P-Tau181), and the 42 amino acid peptide, alpha-amyloid isoform (Aβ) are well established as markers present in cerebrospinal fluid (CSF). However, there is a constant need for new diagnostic markers to identify the disease at a very early stage [87]. A review about the role of proteoforms in the pathophysiology process of Alzheimer's disease was described in [88]. The mass spectrometry performance of three canonical proteins, clusterin, secretogranin-2, or chromogranin A, was presented. Variations on the levels of Apo A-1, a protein with antioxidant and anti-inflammatory properties, in the serum or in CSF, are also indicated as a potential marker for AD diagnosis and progression. Apo A-1 exhibits [86] and inhibits the aggregation and neurotoxicity of an amyloid-β peptide in AD [89]. The possible association between apolipoproteins increased Apo A-1 levels that were correlated with decreasing risk of dementia [87], raising the possibility of a novel role of Apo A-1 in protection against neurological disorders [87, 89].

5.2 Apolipoprotein and lipid metabolism

Possible correlations between apolipoprotein levels (Apo C-III, Apo C-I and Apo C-II) with dyslipidemia and cardiovascular disease were presented in [86]. Apolipoproteins function as the structural components of lipoprotein particles, cofactors for enzymes and ligands for cell-surface receptors. Apolipoproteins exhibit proteoforms associated with nucleotide polymorphisms (SNPs) and post-translational modifications such as glycosylation, oxidation and sequence trunked [86]. The human apo Cs are protein constituents of chylomicrons, VLDL and

HDL. The protein APO C-III has 79 amino acids and can be glycosylated in the residue of Threonine 48. Initially, four APO C-III isoforms were identified by mass spectrometry and later 12 proteoforms. These proteoforms differ by absence of glycosylation (APO C-III Oa), glycosylation (APO C-III Ob), addition of one or two sialic acid residues (APO C-III 1, APO C-III 2) or addition of fucose at glycosylation sites. There are also truncated proteoforms due to amino acid substitution. Increases in APO C-III2 levels are associated with a reduction in TG and LDL levels, and perhaps this is a possible mechanism for dyslipidemia processes and reduced risk of cardiovascular disease (CVD) [86].

5.3 Cancer disease

The identification of novel biomarkers for early clinical-stage cancer detection, targeted molecular therapies, disease monitoring and drug development could impact on the future care of cancer patients. A systematic study of cancer samples using omics technologies, oncoproteomics, is in progress. He et al. summarize the advantages and limitations of the critical technologies used in (onco)proteogenomics [90]. In other studies, Zhan et al. [91] compared MALDI-MS, LC-Q-TOF MS and LC-Orbitrap Velos MS for the identification of proteins within one spot. They described the importance of the development of stable isotope labeling coupled with 2DE-LC/MS in a large-scale study of human proteoforms. This powerful technique platform identified in Blue-stained 2DE spots at least 42 and 63 proteins/spot in an analysis of a human glioblastoma proteome and a human pituitary adenoma proteome, respectively. A critical study to detect new proteomic markers of medullary thyroid carcinoma, combining MALDI-MSI and nLC-ESI-MS/MS were developed by [92]. They identified proteins as moesin, veriscan and lumican and intratumoural amyloid components, including calcitonin, apolipoprotein E, apolipoprotein IV and vitronectin with a potential role in medullary thyroid carcinoma pathogenesis [92].

6. Conclusion

In conclusion, the proteoform identification using a proteomic approach can be an advance in diagnostic routines and development of precision/personalized medicine. Efforts should be concentrated on clinical studies and then on, and one aspect that precludes is the cost and complexity of these tests. Therefore, studies to simplify sample preparation steps and MS platforms need to be performed to reduce cost per test.

Conflict of interest

The authors declare no conflict of interest.

Author details

Jucélia da Silva Araújo and Olga Lima Tavares Machado*
Universidade Estadual do Norte Fluminense-Darcy Ribeiro, Rio de Janeiro, Brazil

*Address all correspondence to: olga@uenf.br

IntechOpen

References

[1] Schlüter H, Apweiler R, Holzhütter HG, Jungblut PR. Finding one's way in proteomics: A protein species nomenclature. Chemistry Central Journal. 2009;3(1):3-11

[2] Larance M, Lamond AI. Multidimensional proteomics for cell biology. Nature Reviews Molecular Cell Biology. 2015;16(5):269-280

[3] Dakna M, He Z, Yu WC, Mischak H, Kolch W. Technical, bioinformatical and statistical aspects of liquid chromatography–mass spectrometry (LC–MS) and capillary electrophoresis-mass spectrometry (CE-MS) based clinical proteomics: A critical assessment. Journal of Chromatography B. 2009;877(13):1250-1258

[4] Zhu W, Smith JW, Huang CM. Mass spectrometry-based label-free quantitative proteomics. BioMed Research International. 2009

[5] Zhang Y, Fonslow BR, Shan B, Baek MC, Yates JR III. Protein analysis by shotgun/bottom-up proteomics. Chemical Reviews. 2013;113(4):2343-2394

[6] Catherman AD, Skinner OS, Kelleher NL. Top down proteomics: Facts and perspectives. Biochemical and Biophysical Research Communications. 2014;445(4):683-693

[7] Dams M, Dores-Sousa JL, Lamers RJ, Treumann A, Eeltink S. High-resolution nano-liquid chromatography with tandem mass spectrometric detection for the bottom-up analysis of complex proteomic samples. Chromatographia. 2019;82(1):101-110

[8] Shortreed MR, Frey BL, Scalf M, Knoener RA, Cesnik AJ, Smith LM. Elucidating proteoform families from proteoform intact-mass and lysine-count measurements.

Journal of Proteome Research. 2016;15(4):1213-1221

[9] Smith LM, Kelleher NL, Kelleher NL. Proteoform: A single term describing protein complexity. Nature Methods. 2013;10(3):186-187. DOI: 10.1038/nmeth.2369

[10] Schaffer LV, Shortreed MR, Cesnik AJ, Frey BL, Solntsev SK, Scalf M, et al. Expanding proteoform identifications in top-down proteomic analyses by constructing proteoform families. Analytical Chemistry. 2017;90(2):1325-1333

[11] Jungblut PR. The proteomics quantification dilemma. Journal of Proteomics. 2014;107:98-102

[12] Jungblut PR, Thiede B, Schlüter H. Towards deciphering proteomes via the proteoform, protein speciation, moonlighting and protein code concepts. Journal of Proteomics. 2016;134:1-4

[13] Schachner LF, Ives AN, McGee JP, Melani RD, Kafader JO, Compton PD, et al. Standard proteoforms and their complexes for native mass spectrometry. Journal of the American Society for Mass Spectrometry. 2019:1-9

[14] Chen B, Peng Y, Valeja SG, Xiu L, Alpert AJ, Ge Y. Online hydrophobic interaction chromatography–mass spectrometry for top-down proteomics. Analytical Chemistry. 2016;88(3):1885-1891

[15] Skinner OS, Havugimana PC, Haverland NA, Fornelli L, Early BP, Greer JB, et al. An informatic framework for decoding protein complexes by top-down mass spectrometry. Nature Methods. 2016;13(3):237

[16] Melani RD, Nogueira FC, Domont GB. It is time for top-down

venomics. Journal of Venomous Animals and Toxins Including Tropical Diseases. 2017;**23**(1):44

[17] Haverland NA, Skinner OS, Fellers RT, Tariq AA, Early BP, LeDuc RD, et al. Defining gas-phase fragmentation propensities of intact proteins during native top-down mass spectrometry. Journal of the American Society for Mass Spectrometry. 2017;**28**(6):1203-1215

[18] Li H, Wolff JJ, Van Orden SL, Loo JA. Native top-down electrospray ionization-mass spectrometry of 158 kDa protein complex by high-resolution Fourier transform ion cyclotron resonance mass spectrometry. Analytical Chemistry. 2013;**86**(1):317-320

[19] Leney AC, Heck AJ. Native mass spectrometry: What is in the name? Journal of the American Society for Mass Spectrometry. 2017;**28**(1):5-13

[20] Schaffer LV, Millikin RJ, Miller RM, Anderson LC, Fellers RT, Ge Y, et al. Identification and quantification of proteoforms by mass spectrometry. Proteomics. 2019;**19**(10):1800361

[21] Shen X, Kou Q, Guo R, Yang Z, Chen D, Liu X, et al. Native proteomics in discovery mode using size-exclusion chromatography–capillary zone electrophoresis–tandem mass spectrometry. Analytical Chemistry. 2018;**90**(17):10095-10099

[22] Skinner OS, Do Vale LH, Catherman AD, Havugimana PC, Sousa MVD, Compton PD, et al. Native GELFrEE: A new separation technique for biomolecular assemblies. Analytical Chemistry. 2015;**87**(5):3032-3038

[23] Capriotti AL, Cavaliere C, Foglia P, Samperi R, Laganà A. Intact protein separation by chromatographic and/or electrophoretic techniques for top-down proteomics.

Journal of Chromatography A. 2011;**1218**(49):8760-8776

[24] Jungblut PR, Holzhütter HG, Apweiler R, Schlüter H. The speciation of the proteome. Chemistry Central Journal. 2008;**2**(16):1-10

[25] Chen B, Brown KA, Lin Z, Ge Y. Top-down proteomics: Ready for prime time? Analytical Chemistry. 2017;**90**(1):110-127

[26] Naryzhny S. Inventory of proteoforms as a current challenge of proteomics: Some technical aspects. Journal of Proteomics. 2019;**191**:22-28

[27] Xiu L, Valeja SG, Alpert AJ, Jin S, Ge Y. Effective protein separation by coupling hydrophobic interaction and reverse phase chromatography for top-down proteomics. Analytical Chemistry. 2014;**86**(15):7899-7906

[28] Wessels HJ, Vogel RO, Lightowlers RN, Spelbrink JN, Rodenburg RJ, van den Heuvel LP, et al. Analysis of 953 human proteins from a mitochondrial HEK293 fraction by complexome profiling. PLoS One. 2013;**8**(7):e68340

[29] Ishii K, Zhou M, Uchiyama S. Native mass spectrometry for understanding dynamic protein complex. Biochimica et Biophysica Acta (BBA)-General Subjects. 2018;**1862**(2):275-286

[30] Donnelly DP, Rawlins CM, DeHart CJ, Fornelli L, Schachner LF, Lin Z, et al. Best practices and benchmarks for intact protein analysis for top-down mass spectrometry. Nature Methods. 2019;**16**(7):587

[31] O'Farrell PH. High resolution two-dimensional electrophoresis of proteins. Journal of Biological Chemistry. 1975;**250**:407-421

[32] Issaq HJ, Veenstra TD. Two-dimensional polyacrylamide gel

electrophoresis (2D-PAGE): Advances and perspectives. BioTechniques. 2008;**44**(5):697-700

[33] Westermeier R. Looking at proteins from two dimensions: A review on five decades of 2D electrophoresis. Archives of Physiology and Biochemistry. 2014;**120**(5):168-172

[34] Bjellqvist B, Ek K, Righetti PG, Gianazza E, Görg A, Westermeier R, et al. Isoelectric focusing in immobilized pH gradients: Principle, methodology and some applications. Journal of Biochemical and Biophysical Methods. 1982;**6**(4):317-339

[35] Taylor RC, Coorssen JR. Proteome resolution by two-dimensional gel electrophoresis varies with the commercial source of IPG strips. Journal of Proteome Research. 2006;**5**(11):2919-2927

[36] Padula M, Berry I, Raymond B, Santos J, Djordjevic SP. A comprehensive guide for performing sample preparation and top-down protein analysis. Proteomes. 2017;**5**(2):11

[37] Arentz G, Weiland F, Oehler MK, Hoffmann P. State of the art of 2D DIGE. PROTEOMICS–Clinical Applications. 2015;**9**(3-4):277-288

[38] Collier TS, Muddiman DC. Analytical strategies for the global quantification of intact proteins. Amino Acids. 2012;**43**(3):1109-1117

[39] Marouga R, David S, Hawkins E. The development of the DIGE system: 2D fluorescence difference gel analysis technology. Analytical and Bioanalytical Chemistry. 2005;**382**(3):669-678

[40] Kirana C, Peng L, Miller R, Keating JP, Glenn C, Shi H, et al. Combination of laser microdissection, 2D-DIGE and MALDI-TOF MS to identify protein biomarkers to predict colorectal cancer sprad. Clinical Proteomics. 2019;**16**(1):3

[41] Tran JC, Doucette AA. Gel-eluted liquid fraction entrapment electrophoresis: An electrophoretic method for broad molecular weight range proteome separation. Analytical Chemistry. 2008;**80**(5):1568-1573

[42] Kim KH, Compton PD, Tran JC, Kelleher NL. Online matrix removal platform for coupling gel-based separations to whole protein electrospray ionization mass spectrometry. Journal of Proteome Research. 2015;**14**(5):2199-2206

[43] Wessel DM, Flügge UI. A method for the quantitative recovery of protein in dilute solution in the presence of detergents and lipids. Analytical Biochemistry. 1984;**138**(1):141-143

[44] Armirotti A, Damonte G. Achievements and perspectives of top-down proteomics. Proteomics. 2010;**10**(20):3566-3576

[45] Kellie JF, Tran JC, Lee JE, Ahlf DR, Thomas HM, Ntai I, et al. The emerging process of top down mass spectrometry for protein analysis: Biomarkers, protein-therapeutics, and achieving high throughput. Molecular BioSystems. 2010;**6**(9):1532-1539

[46] Melani RD, Skinner OS, Fornelli L, Domont GB, Compton PD, Kelleher NL. Mapping proteoforms and protein complexes from king cobra venom using both denaturing and native top-down proteomics. Molecular & Cellular Proteomics. 2016;**15**(7):2423-2434

[47] Vellaichamy A, Tran JC, Catherman AD, Lee JE, Kellie JF, Sweet SM, et al. Size-sorting combined with improved nanocapillary liquid chromatography-mass spectrometry for identification of intact proteins up to 80 kDa. Analytical Chemistry. 2010;**82**(4):1234-1244

[48] Kohl FJ, Sánchez-Hernández L, Neusuess C. Capillary electrophoresis in two-dimensional separation systems: Techniques and applications. Electrophoresis. 2015;**36**(1):144-158

[49] Han X, Wang Y, Aslanian A, Bern M, Lavallée-Adam M, Yates JR III. Sheathless capillary electrophoresis-tandem mass spectrometry for top-down characterization of pyrococcus furiosus proteins on a proteome scale. Analytical Chemistry. 2014;**86**(22):11006-11012

[50] McCool EN, Chen D, Li W, Liu Y, Sun L. Capillary zone electrophoresis-tandem mass spectrometry with ultraviolet photodissociation (213 nm) for large-scale top-down proteomics. Analytical Methods. 2019;**11**(22):2855-2861

[51] Chen D, Shen X, Sun L. Capillary zone electrophoresis-mass spectrometry with microliter-scale loading capacity, 140 min separation window and high peak capacity for bottom-up proteomics. Analyst. 2017;**142**(12):2118-2127

[52] Chen B, Lin Z, Alpert AJ, Fu C, Zhang Q, Pritts WA, et al. Online hydrophobic interaction chromatography–mass spectrometry for the analysis of intact monoclonal antibodies. Analytical Chemistry. 2018;**90**(12):7135-7138

[53] Camerini S, Mauri P. The role of protein and peptide separation before mass spectrometry analysis in clinical proteomics. Journal of Chromatography A. 2015;**1381**:1-12

[54] Cai W, Tucholski TM, Gregorich ZR, Ge Y. Top-down proteomics: Technology advancements and applications to heart diseases. Expert Review of Proteomics. 2016;**13**(8):717-730

[55] Wang T, Chen D, Lubeckyj RA, Shen X, Yang Z, McCool EN, et al.

Capillary zone electrophoresis-tandem mass spectrometry for top-down proteomics using attapulgite nanoparticles functionalized separation capillaries. Talanta. 2019;**202**:165-170

[56] Zhang Z, Wu S, Stenoien DL, Paša-Tolić L. High-throughput proteomics. Annual Review of Analytical Chemistry. 2014;**7**:427-454

[57] Cavaliere C, Capriotti A, La Barbera G, Montone C, Piovesana S, Laganà A. Liquid chromatographic strategies for separation of bioactive compounds in food matrices. Molecules. 2018;**23**(12):3091

[58] Wang Z, Ma H, Smith K, Wu S. Two-dimensional separation using high-pH and low-pH reversed phase liquid chromatography for top-down proteomics. International Journal of Mass Spectrometry. 2018;**427**:43-51

[59] Valeja SG, Xiu L, Gregorich ZR, Guner H, Jin S, Ge Y. Three dimensional liquid chromatography coupling ion exchange chromatography/hydrophobic interaction chromatography/reverse phase chromatography for effective protein separation in top-down proteomics. Analytical Chemistry. 2015;**87**(10):5363-5371

[60] Patrie SM. Top-down mass spectrometry: Proteomics to proteoforms. Nature Methods. 2016;**4**(9):709. In Modern Proteomics–Sample Preparation, Analysis and Practical Applications (pp. 171-200). Springer, Cham. analysis

[61] Tassi M, De Vos J, Chatterjee S, Sobott F, Bones J, Eeltink S. Advances in native high-performance liquid chromatography and intact mass spectrometry for the characterization of biopharmaceutical products. Journal of Separation Science. 2018;**41**(1):125-144

[62] Huang TY, Chi LM, Chien KY. Size-exclusion chromatography using

reverse-phase columns for protein separation. Journal of Chromatography A. 2018;**1571**:201-212

[63] Cai W, Tucholski T, Chen B, Alpert AJ, McIlwain S, Kohmoto T, et al. Top-down proteomics of large proteins up to 223 kDa enabled by serial size exclusion chromatography strategy. Analytical Chemistry. 2017;**89**(10):5467-5475

[64] Gargano AF, Roca LS, Fellers RT, Bocxe M, Domínguez-Vega E, Somsen GW. Capillary HILIC-MS: A new tool for sensitive top-down proteomics. Analytical Chemistry. 2018;**90**(11):6601-6609

[65] Tengattini S, Domínguez-Vega E, Temporini C, Bavaro T, Rinaldi F, Piubelli L, et al. Hydrophilic interaction liquid chromatography-mass spectrometry as a new tool for the characterization of intact semi-synthetic glycoproteins. Analytica Chimica Acta. 2017;**981**:94-105

[66] Periat A, Krull IS, Guillarme D. Applications of hydrophilic interaction chromatography to amino acids, peptides, and proteins. Journal of Separation Science. 2015;**38**(3):357-367

[67] Periat A, Fekete S, Cusumano A, Veuthey JL, Beck A, Lauber M, et al. Potential of hydrophilic interaction chromatography for the analytical characterization of protein biopharmaceuticals. Journal of Chromatography A. 2016;**1448**:81-92

[68] Regnier FE, Kim J. Proteins and proteoforms: New separation challenges. Analytical Chemistry. 2017;**90**(1):361-373

[69] D'Atri V, Fekete S, Beck A, Lauber M, Guillarme D. Hydrophilic interaction chromatography hyphenated with mass spectrometry: A powerful analytical tool for the comparison of originator and biosimilar therapeutic monoclonal antibodies at the middle-up level of analysis. Analytical Chemistry. 2017;**89**(3):2086-2092

[70] Gargano AF, Shaw JB, Zhou M, Wilkins CS, Fillmore TL, Moore RJ, et al. Increasing the separation capacity of intact histone proteoforms chromatography coupling online weak cation exchange-HILIC to reversed phase LC UVPD-HRMS. Journal of Proteome Research. 2018;**17**(11):3791-3800

[71] Tian Z, Tolić N, Zhao R, Moore RJ, Hengel SM, Robinson EW, et al. Enhanced top-down characterization of histone post-translational modifications. Genome Biology. 2012;**13**(10):R86

[72] Cole J, Hanson EJ, James DC, Dockrell DH, Dickman MJ. Comparison of data-acquisition methods for the identification and quantification of histone post-translational modifications on a Q Exactive HF hybrid quadrupole Orbitrap mass spectrometer. Rapid Communications in Mass Spectrometry. 2019;**33**(10):897-906

[73] Monti C, Zilocchi M, Colugnat I, Alberio T. Proteomics turns functional. Journal of Proteomics. 2019;**198**:36-44

[74] Yuan ZF, Arnaudo AM, Garcia BA. Mass spectrometric analysis of histone proteoforms. Annual Review of Analytical Chemistry. 2014;**7**:113-128

[75] Toby TK, Fornelli L, Kelleher NL. Progress in top-down proteomics and the analysis of proteoforms. Annual Review of Analytical Chemistry. 2016;**9**:499-519

[76] Koopmans F, Ho JT, Smit AB, Li KW. Comparative analyses of data independent acquisition mass spectrometric approaches: DIA, WiSIM-DIA, and untargeted DIA. Proteomics. 2018;**18**(1):1700304

[77] Mann M, Hendrickson RC, Pandey A. Analysis of proteins and proteomes by mass spectrometry. Annual Review of Biochemistry. 2001;**70**(1):437-473

[78] van Belkum A, Welker M, Erhard M, Chatellier S. Biomedical mass spectrometry in today's and tomorrow's clinical microbiology laboratories. Journal of Clinical Microbiology. 2012;**50**(5):1513-1517

[79] Nicolardi S, Switzar L, Deelder AM, Palmblad M, van der Burgt YE. Top-down MALDI-in-source decay-FTICR mass spectrometry of isotopically resolved proteins. Analytical Chemistry. 2015;**87**(6):3429-3437

[80] Naryzhny SN, Zgoda VG, Maynskova MA, Novikova SE, Ronzhina NL, Vakhrushev IV, et al. Combination of virtual and experimental 2DE together with ESI LC-MS/MS gives a clearer view about proteomes of human cells and plasma. Electrophoresis. 2016;**37**(2):302-309

[81] Ahlf DR, Compton PD, Tran JC, Early BP, Thomas PM, Kelleher NL. Evaluation of the compact high-field orbitrap for top-down proteomics of human cells. Journal of Proteome Research. 2012;**11**(8):4308-4314

[82] Tucholski T, Knott SJ, Chen B, Pistono P, Lin Z, Ge Y. A top-down proteomics platform coupling serial size exclusion chromatography and Fourier transform ion cyclotron resonance mass spectrometry. Analytical Chemistry. 2019;**91**(6):3835-3844

[83] Olsen JV, Macek B, Lange O, Makarov A, Horning S, Mann M. Higher-Energy C-Trap Dissociation for Peptide Modification; 2007

[84] Zhang H, Ge Y. Comprehensive analysis of protein modifications by top-down mass spectrometry.

Circulation: Cardiovascular Genetics. 2011;**4**(6):711-711

[85] Fekete S, Veuthey JL, Beck A, Guillarme D. Hydrophobic interaction chromatography for the characterization of monoclonal antibodies and related products. Journal of Pharmaceutical and Biomedical Analysis. 2016;**130**:3-18

[86] Nedelkov D. Human proteoforms as new targets for clinical mass spectrometry protein tests. Expert Review of Proteomics. 2017;**14**(8):691-699. DOI: 10.1080/14789450.2017.1362337

[87] Fania C, Arosio B, Capitanio D, Torretta E, Gussago C, Ferri E, et al. Protein signature in cerebrospinal fluid and serum of Alzheimer's disease patients: The case of apolipoprotein A-1 proteoforms. PLoS One. 2017;**12**(6):e0179280. DOI: 10.1371/journal.pone.0179280

[88] Vialaret J, Schmit PO, Lehmann S, Gabelle AJ, Bern M, Paape R, et al. Identification of multiple proteoforms biomarkers on clinical samples by routine Top-Down approaches. Data in Brief. 2018;**18**:1013-1021. DOI: 10.1016/j.dib.2018.03.114

[89] Naveed M, Mubeen L, Khan A, Ibrahim S, Meer B. Plasma biomarkers: Potent screeners of Alzheimer's disease. American Journal of Alzheimer's Disease & Other Dementias. 2019;**34**(5):290-301

[90] He Y, Mohamedali A, Canhua Huang C, Baker MS, Nice EC. Oncoproteomics: Current status and future opportunities. Clinica Chimica Acta. 2019;**495**:611-624

[91] Zhan X, Yang H, Peng F, Li J, Mu Y, Long Y, et al. How many proteins can be identified in a 2DE gel spot within an analysis of a complex human cancer tissue proteome? Electrophoresis. 2018;**39**:965-980

[92] Smith A, Galli M, Piga I, Denti V, Stella M, Chinello C, et al. Molecular signatures of medullary thyroid carcinoma by matrix-assisted laser desorption/ionisation mass spectrometry imaging. Journal of Proteomics. 2019;**16**:114-123

Chapter 3

Preparing Proteoforms of Therapeutic Proteins for Top-Down Mass Spectrometry

Siti Nurul Hidayah, Manasi Gaikwad, Laura Heikaus and Hartmut Schlüter

Abstract

A characteristic of many proteoforms, derived from a single gene, is their similarity regarding the composition of atoms, making their analysis very challenging. Many overexpressed recombinant proteins are strongly associated with this problem, especially recombinant therapeutic glycoproteins from large-scale productions. In contrast to small molecule drugs, which consist of a single defined molecule, therapeutic protein preparations are heterogenous mixtures of dozens or even hundreds of very similar species. With mass spectrometry, currently high-quality spectra of intact proteoforms can be obtained only, if the complexity of the mixture of individual proteoform-ions, entering the gas phase at the same time is low. Thus, prior to mass spectrometric analysis, an effective separation is required for getting fractions with a low number of individual proteoforms. This is especially true not only for recombinant therapeutic proteins, because of their huge heterogeneity, but also relevant for top-down proteomics. Purification of proteoforms is the bottleneck in analyzing intact proteoforms with mass spectrometry. This review is focusing on the current state of the art, especially of liquid chromatography for preparing proteoforms for mass spectrometric top-down analysis. The topic of therapeutic proteins has been chosen, because this group of proteins is most challenging regarding their proteoform analysis.

Keywords: proteoforms, top-down mass spectrometry, therapeutic proteins, liquid chromatography, protein purification parameter screening, displacement chromatography

1. Introduction

The analysis of proteoforms, often also termed protein species or isoforms, is the next level in proteomics. The first comprehensive definition of this subgroup of proteins was published by Jungblut et al. [1] and Schlüter et al. [2], using the term "protein species". In 2013, Smith and Kelleher [3] introduced the term "proteoform", which today is widely accepted in the community of proteomics experts. The concept of "proteoform" is nearly identical with the concept of "protein species". The only difference is that the proteoform concept is gene-centric and the protein-species-concept is chemistry-centric.

For developing methods for comprehensive analysis of proteoforms, the group of therapeutic proteins is a suitable training area. Therapeutic proteins are known to be rich in the number of proteoforms. Although a therapeutic protein product is containing only trace amounts of impurities like host cell proteins, which are difficult to detect because of their very low concentration, the analysis of their proteoforms is very challenging because of their large number, their similarity and their low concentration compared to the main proteoform.

2. Analysis of proteoforms: challenges

The most common method in proteomics is the bottom-up or shotgun approach. It relies on the proteolytic cleavage of proteins by proteases like trypsin. The resulting peptide mixture is subjected to liquid chromatography coupled to tandem mass spectrometry (LC–MS/MS) analysis. Proteins are identified from the LC–MS/MS data by comparing the peptide fragment spectra against in-silico fragment spectra generated from a protein database [4]. As a rule of thumb, a protein is claimed to be identified, if at least two unique peptides are identified representing parts of the sequence. Thus, often a sequence coverage of 100% is not obtained. Consequently, if this is the case, it can be only stated that a product or several products (proteoforms) of a defined gene has been identified. No information about the identity of the underlying proteoform is obtained. It can even be assumed that the identified tryptic peptides may be products of several different proteoforms. For the characterization of a therapeutic protein, bottom-up proteomics is a standard method. The signals in the LC–MS chromatograms represent tryptic peptides of all proteoforms of the therapeutic protein. A defined tryptic peptide, which is present in all proteoforms, will form one single monoisotopic signal. Its signal intensity represents the sum of this peptide from the different species. The presence of an individual proteoform only can be detected, if this proteoform will yield a tryptic peptide, a defined phosphor-peptide, which is unique for this proteoform. However, it cannot be excluded, that there are several proteoforms containing that peptide. As a result, bottom-up proteomics is helpful for getting LC–MS chromatograms which can be used as fingerprints of a therapeutic protein, but will give no information about the number and composition of proteoforms within the therapeutic protein product. The detection of a low abundant proteoform is especially difficult, since a unique tryptic peptide of such a proteoform is present in a low amount and thereby the signal in a bottom-up proteomics LC–MS chromatogram will have a low intensity. Thus, if the detection of different proteoforms is of interest, top-down mass spectrometry (TDMS) is the method of choice, because it utilizes the intact proteoform for analysis instead of proteolytic peptides.

For performing a TDMS analysis, a purified individual intact proteoform is transferred into the MS. From the MS spectrum of the intact ions, the molecular weight can be determined. Various techniques are available for fragmentation of the intact proteoform such as HCD, CID, ETD, ETHcD, ECD, UVPD and IRMPD, yielding different types for fragments, which complement each other [5]. After fragmentation, the proteoform can be identified by interpreting the fragment spectrum. There are several software tools available for analyzing the TDMS intact data [6–8]. The review of Schaffer et al. is recommended as an introduction into TDMS [9]. Robust protocols for mass analysis of intact proteins with TDMS were recently published by Donnelly et al. [10]. TDMS is requiring sample mixtures of low complexity for obtaining high quality spectra of proteoforms. Aebersold et al. estimated the number of proteoforms being present in the human organism in the range of approximately a billion [11]. Thus, very efficient purification steps prior to the TDMS are required to tackle the huge number of individual proteoforms in cells and tissues of body fluids. Beside the excessive number of individual proteoforms, their dynamic range is a further challenge.

3. Analysis of proteoforms of recombinant therapeutic proteins: challenges

Similar challenges are associated with recombinant therapeutic proteins. The importance of therapeutic proteins has been continually increasing over the past years [12, 13]. Currently, several types of therapeutic proteins [14] are available in the market including monoclonal antibodies (mAbs), erythropoietin (EPO), insulin, human growth hormone and many more. Therapeutic proteins market is dominated by the monoclonal antibodies with sales of approximately $123 billion in 2017 and will be seen increasing with the upcoming biosimilar market [13]. Therapeutic proteins possess several advantages over small molecule drugs due to their higher specificity towards drug targets, which are in most cases also proteins [15]. This makes therapeutic proteins able to target specific key steps in disease pathology [16].

This group of man-made proteins has presumably a significantly higher number of proteoforms per gene than proteoforms per gene *in vivo*, causing a huge number of proteoforms within a single recombinant therapeutic protein (rTP) product. The heterogeneity is developing during the production of an rTP mainly in the upstream processing. The first event increasing the heterogeneity is alternative splicing [17–19]. The second critical step is the protein biosynthesis at the ribosomes, in which errors can occur. Proteolytic cleavage may happen at any stage after the protein has left the ribosome, not only within the host cell, but also extracellularly, if host cell proteases have not been removed by purification of the target protein. Many therapeutic proteins like conventional monoclonal antibodies or erythropoietin [20] are posttranslationally modified by glycans. Especially, the glycan chains are adding an additional factor multiplying the heterogeneity of proteoforms. An example of a therapeutic glycoprotein is Etanercept, which is decorated with O- and N-glycans. Commercial preparations of Etanercept used as drugs show a very high degree of complexity [21]. It can be assumed that therapeutic fusion proteins applied to patients like etanercept are containing even hundreds of species, which differ in their exact composition of atoms. In addition to glycans, all other forms of posttranslational modifications are possible, depending on the nature of the protein and the type of the host cells and the upstream parameters.

Why is the heterogeneity of recombinant therapeutic proteins much higher than the heterogeneity of gene products in-vivo? Host cells used for the production of recombinant therapeutic proteins are optimized to synthesize a large excess of recombinant proteins [22]. However, increasing the expression of proteins does not usually correlate to increase in the correctly processed bioactive form of the recombinant proteins [22]. Consequently, the probability is increasing, that these overexpressed recombinant proteins are underlying errors during synthesis, side reactions of enzymes and spontaneous chemical reactions. As a result, the number of recombinant species, which have a low quality, is much higher than in a native cell in an intact organism [23]. It was reported that overexpressing recombinant therapeutic proteins is also accompanied by an increase in high molecular weight aggregates and misfolded forms [24]. Thus, it can be assumed that the cellular systems, which usually remove low-quality or incorrectly processed proteins, are swamped by these inadequate proteins [25] and thereby these species will not be processed in the cell or be eliminated. Beside the enzymatic reactions mainly taking place in the upstream-processing, chemical reactions which modify the recombinant therapeutic proteins, can occur during the whole production process including even the final product fill and finish or storage [26, 27]. A very common reaction is the oxidation of methionine, which can happen on nearly every stage of the production and can affect the efficacy of the product.

Is any risk associated with the large number of species? Fortunately, severe side effects associated with species, which are not exactly identical with the target protein, have been reported very seldomly. An unfortunate case with dramatic consequences for a few patients was reported from Seidl et al. [28]. In this case, tungsten ions, a contamination which got into the glass vials during the production of the vials, induced the dimerization of erythropoietin. As a result, a few patients developed autoantibodies against erythropoietin, thereby destroying the remaining cells in these patients, which were producing the native hormone. Since a therapy with erythropoietin was not possible any more, these patients had to get blood transfusions for survival. Non-human glycan structures bound to therapeutic proteins, which can occur when producing them in mouse cells, can induce hypersensitivity reactions [29, 30].

More common than severe side effects is the phenomenon that , showing even small differences in their composition of atoms compared with the target species, make the species less potent than the target species. For example, deamidation, causing a + 1 Da shift of the molecular weight, can decrease the efficacy of a therapeutic protein [31], as observed with recombinant human interleukin (rhIL)-15 [32]. Deamidation converts asparagine or glutamine to aspartic acid or glutamic acid, respectively. As a result, the polar, uncharged amides are changed into negatively charged carboxylic acids, impacting protein surface-charge density and surface hydrophobicity, thereby explaining the change of the efficacy of a therapeutic protein. Deamidation of asparagine can occur spontaneously at physiological pH of 7.4 [32]. A further important modification of proteins is the disulfide bond (S-S), which is formed by the oxidation of thiol groups (SH) between two cysteine residues resulting in a covalent bond [33], which is decreasing the molecular weight of a protein by 2 Da. Disulfide-bonds have an impact on protein stability as well as on activities [33]. Du et al. stated that during the manufacturing process, extensive reduction of antibodies has been observed after harvest operation or Protein A affinity chromatography and multiple process parameters correlate to the extent of the reduction [34]. The topic "disulfide bonds of therapeutic proteins" is in depth discussed by Lakbub et al. [35].

More details about sources and effects of microheterogeneity are described in the excellent reviews of Beyer [36] and Ambrogelly [37].

How large are the differences of the individual proteoforms of a therapeutic protein? Proteofroms can vary in all chemical properties known, such as size, isoelectric points (pI) [38] and hydrophobicity [39]. The pIs of recombinant erythropoietin varies from pH 3.5–6 [38, 40]. Therapeutic proteins are characterized by the presence of size variants arising from the manufacturing process or storage conditions when exposed to chemical, physical or conformational stress [41]. These size variants may include the N terminus clipped proteins, truncated forms, fragments representing sub molecular weight species or improperly assembled therapeutic proteins. The formation of dimers or multimers, in which more than two monomers are forming a complex, is a problem, which many therapeutic proteins are associated with [42]. Such aggregates can induce adverse immune responses in patients [43]. The proteoforms of recombinant erythropoietin are varying within a range of 4–6 kDa [20]. Beside these larger differences in size, the composition of atoms of many proteoforms derived from one single gene can be very similar within subtypes of proteoforms such as the family of acidic proteoforms. As a result, the separation of charge variants by ion exchange usually is successful but the composition within a single fraction might not only contain one single but also multiple proteoforms [44].

4. Separation of proteoforms of therapeutic proteins

4.1 Separation of proteoforms of therapeutic proteins with liquid chromatography

Liquid chromatography (LC) is the most common for purification and fractionation of therapeutic proteins [37]. The proteoforms are either separated by size-exclusion (SEC), making use of different path lengths through chromatographic particles related to the size of the proteins, or by adsorption chromatography. The latter is applying the principle of separation of molecules by their different velocities during crossing a column filled with chromatographic particles. The velocities are proportional to the affinities of the molecules towards the stationary phase of the stationary phase. Depending on the chemistry of the functional groups of the stationary phase, different forms of liquid chromatography are possible based on adsorption to the stationary phase, highlighted in bold in **Table 1**.

Table 1 is giving an overview about the different types of separation methods and their frequency of application with a focus on therapeutic proteins and in addition with respect to proteoforms. The numbers of column 2 compared with column 3 clearly show that the topic of proteoforms is not yet addressed very often. The selected reviews will give deeper insights into the different separation methods.

Affinity chromatography using chromatographic material derivatized with protein-A is the most common and effective method for the purification of recombinant monoclonal antibodies [45]. For the separation of proteoforms of recombinant monoclonal antibodies, it is not very relevant.

Ion exchange chromatography (IEX): charge variants of therapeutic proteins such as acidic or basic species can be separated with ion exchange chromatography (IEX) [46]. IEX of proteins can be performed with oppositely charged ionic group on the stationary phase as either anion exchange or cation exchange chromatography. Elution buffers are decreasing electrostatic interactions of the proteins with IEX material thereby decreasing the affinity of the protein towards the stationary phase. Elution can be either pH or salt based [47]. Salt-based elution is used for IEX with ultra violet (UV) online detection. Coupling IEX directly with MS is only possible if the elution buffer system is volatile [48]. Acidic species are often related to PTM's like sialic acid or deamidation on asparagine, while basic variants are formed by aspartate isomerization, succinimide formation, variants of C terminal lysine and N terminal glutamine [49]. IEX is giving relative quantitative information about charge variants which can be important for the qualification of manufacturing batches [50].

Hydroxyapatite-chromatography (HAP) is based on a material consisting of the crystals of calcium hydroxyapatite, described by the formula $Ca_5(PO_4)_3(OH)$. HAP can be described as mixed-mode chromatography. The Ca^{2+} −ions can act via electrostatic interactions as anion-exchanger. Also, metal coordination bonds of carboxylic groups can be formed with the Ca^{2+} −ions. With the anionic phosphate groups of HAP, positive-charged molecules will be adsorbed by electrostatic interactions. Phosphate-, chloride-ion-, and calcium-ion- gradients are common as well as multi-component gradients [39]. Therefore, finding appropriate eluents

Separation method	Hits of the PubMed search: monoclonal recombinant antibody OR therapeutic-proteins OR biotherapeutics AND "name of the separation method: left column"	Hits after adding filter: "AND (isoform OR variant OR species OR proteoform)"	Review
Affinity chromatography (AF)	1046	189	[45, 100]
Anion exchange chromatography (AEC)	80	20	[101, 102]
Cation exchange chromatography (CEX)	56	23	[49, 103]
Hydroxyapatite chromatography (HAP)	6	1	[104]
Hydrophobic interaction chromatography (HIC)	46	10	[39]
Hydrophilic interaction chromatography (HILIC)	8	3	[51, 52]
Immobilized metal affinity chromatography (IMAC)	59	5	[105, 106]
Mixed mode chromatography (MM)	10	0	[59]
Reversed phase chromatography (RP)	170	41	[107]
Size exclusion chromatography (SEC)	973	142	[108]
Liquid chromatography (LC)	1969	308	[109]
Two dimensional liquid chromatography (2D-LC)	11	2	[110, 111]
Capillary electrophoresis (CE)	151	31	[112]
Two dimensional electrophoresis (2DE)	12	4	[38, 113]

The second column is listing the hits got by screening the knowledgebase PubMed (screening date 24.07.2019 8:00 pm) with the search terms: "monoclonal recombinant antibody OR therapeutic-proteins OR biotherapeutics AND anion-exchange-Chromatography". The third column is presenting the hits after adding the filter: "AND (isoform OR variant OR species OR proteoform)". In bold: Forms of liquid chromatography based on adsorption.

Table 1.
Overview about methods for the separation of therapeutic proteins showing the frequency of their application and recommended reviews.

is more difficult than with anion-exchange chromatography. However, screening systematically appropriate parameters of eluent systems should offer the chance to separate proteoforms. As indicated in **Table 1**, HAP is not very often applied for the chromatography of therapeutic proteins, which may be associated with the fact that it is more complex to find optimal elution systems.

Hydrophilic interaction chromatography (HILIC) is making use of high affinities of polar and hydrophilic molecules to hydrophilic stationary phase [51, 52]. Usually the sample application buffer has a high content (>80%) of an organic solvent like acetonitrile. Thus, it is working well for glycans. However, proteins under these conditions may precipitate. If proteoforms will not precipitate, HILIC is an interesting alternative for other forms of adsorption chromatography, especially, if precipitation proteoforms will be removed from the proteoforms of interest.

Hydrophobic interaction chromatography (HIC) is yet another method which can be used for separating different proteoforms of a therapeutic protein. These separations rely on the varying hydrophobicity profiles due to change in conformation of the protein HIC separations use reverse salt gradients and can operate in nondenaturing mode [53]. HIC was presented as a reliable method for monitoring oxidation of tryptophan residues in complement determining region (CDR) of recombinant mAbs [54]. HIC is effective in resolving the proteoforms of antibody drug conjugates varying in drug to antibody ratio [55]. Charge variants coeluting with IEX can be resolved with HIC in the second dimension of separation. Douglas and colleagues demonstrated the separation of carboxy terminal variants, isomerization variants with HIC which could not be resolved at the IEX level [56]. Quantitative information of the succinimide variants was given by HIC with TSKgel butyl-NPR column [57]. Similar application can be also found in detection of impaired disulfide bonding. Typical HIC buffers like ammonium sulfate are requiring desalting of the proteins prior to the MS [58]. Recently, direct coupling of HIC with MS for detailed characterization of mAbs was demonstrated by applying a volatile ammonium acetate buffer [53].

Immobilized metal-affinity-chromatography (IMAC) is widely used for enriching recombinant proteins with histidine tags from a protein extract from host cells. For production of therapeutic proteins, IMAC is not very often used, because metal ions are bleeding into the product. Metal ions like nickel or copper are critical for patients. For the separation of subgroups of proteoforms for analytical purposes also IMAC is an option.

Mixed mode chromatography (MM) is performed with stationary phases which consist of at least two different functional groups [59], like hydroxy apatite (see above). Consequently, a MM material offers two or more types of chromatography. HAP is combining anion exchange (AEX), cation exchange (CEX) and IMAC. Also, with SEC mixed mode chromatography is possible, as described by Schlüter et al. [60]. In that study the electrostatic interaction induced by anionic sugars, which are part of a dextran polymer, were used to separate vanillylmandelic acid, glycine and phenylalanine from each other with a SEC column, which is usually applied for the separation of proteins in the range of 10–100 kDa. Mixed mode chromatography is not very often described for the chromatography of therapeutic proteins (**Table 1**), but it has a huge potential for the separation of proteoforms. For successful separations a rational screening of appropriate parameters is recommended.

Size exclusion chromatography (SEC) is a gold standard for monitoring the presence of aggregates of therapeutic proteins. SEC uses porous stationary phase material wherein the size variants are separated based on the differential access to the pores of the SEC material resulting in different path lengths in relationship to the size [61, 62]. SEC is effectively separating low molecular weight and high molecular weight species in mAbs [63]. SEC has found many applications like stability testing [64], quality control during manufacturing [65], in depth characterization of antibody-drug-conjugates (ADC's) [66] and assessing aggregate content in biosimilarity studies [67]. However, resolution of SEC is rather poor to clearly distinguish individual size variants. Non-specific adsorption to the SEC material can result in peak broadening thereby decreasing resolution. This problem may be minimized by use of organic modifiers in mobile phase or adjusting the pH in relation to the pI of therapeutic protein [61]. Advances in the chemistries of stationary phases incorporating very small core-shell particles or the use of sub-micron particles are improving the resolution of SEC columns [61].

Reversed phase liquid chromatography (RPLC) mainly exploits the differences in hydrophobic properties of molecules for their separation. Sample application onto RPLC columns is performed with eluents having a high content of water, supporting a high affinity of the molecules in the sample towards the stationary phase, which is hydrophobic. Elution is achieved with gradients increasing the concentration of organic solvents in the eluent. Coupling RPLC with the high sensitivity detectors can provide qualitative and quantitative information of the cleaved, modified proteoforms along with main form [68]. Ambrogelly et al. reported RPLC as a method giving a first-hand check of the product quality to help in optimizing the purification strategy [69]. When coupled to high resolution mass spectrometric detection, RPLC also allows distinction of the major glycoforms. More than a decade ago, Dillion presented RPLC not only for determining the intact mAb glycosylation profile but only with the use of high temperature and organic solvents with high eluotropic strength coefficients [70]. Many advancement to conventional RPLC columns have come up in recent times to improve the separation of large therapeutic proteins at milder conditions [71].

The major concern in the use of RPLC for protein separations is the presence of organic solvents, which may precipitate proteins. Since precipitation will occur on the column, it is very difficult to recognize. In the case of proteoforms, it can be assumed that some may be more prone to precipitation than others. As a result, the chromatogram, in which signals from some but not all proteoforms are present, may be misinterpreted since the chromatogram is giving no information about the proteoforms which got lost by precipitation. TDMS protocols often apply RPLC for the analysis of proteoforms, because those species, which elute, are present in a liquid, which is optimal for electrospray ionization (ESI). Because of the problem with precipitation of proteins in RPLC in all TDMS approaches the question is how representative the TDMS chromatogram is regarding the original composition of proteoforms or vice versa how many proteoforms got lost during RPLC.

Elution modes of liquid chromatography: beside the different types of stationary phases, different elution modes are existing, which have an impact on the separation of molecules, namely isocratic elution, gradient elution (GE) and displacement elution (DE). DE is typically using the same sample application buffer and adsorption chromatography materials as gradient elution. In contrast to GE, DE is not using a salt gradient with an increasing concentration of a salt having a low affinity towards the stationary phase, but the elution buffer of DE is consisting of the sample application buffer, into which the displacer is added. The displacer ideally should have an affinity to the stationary phase higher than any of the sample components. After the sample application onto the column is finished, the eluent containing the displacer is immediately pumped onto the column. At the beginning, the displacer molecules are binding strongly to the top of the column, thereby displacing the sample component with the highest affinity. These sample components then displace the sample components with a lower affinity and so on. By this process, bands are formed moving down the column, driven by the displacer. The DE is finished, as soon as the displacer has saturated the stationary phase of the column completely. Within a band a high purity of the component is achieved [72]. DE has been shown to be suitable for separation of complex mixture of tryptic peptides [73–75] and proteins [76–79]. One of the characteristics of DE is that DE has a different selectivity compared with GE [77]. This is one important argument for using DE for the separation of proteoforms. Thus, it is not surprising, that DE has been applied to the separation of proteoforms of therapeutic proteins successfully [46, 80–84].

Rational screening of parameters of liquid chromatography is recommended for optimal results of the separation of proteoforms. The first method describing multi-parallel high-throughput screening for parameters of liquid chromatography was published 2002 by the group of Cramer [85]. In this case, the authors screened for displacers for ion-exchange systems. In the following year the group reported a multi-parallel high-throughput screening for displacers based on batch chromatography [86]. Thiemann et al. published a similar approach termed protein-purification parameter screening system (PPS), which was not focusing on the identification of appropriate displacers but more general on any kind of parameters for adsorption chromatography, independent of the elution mode [87]. The PPS was successfully applied for purification and identification of an angiotensin-II generating enzyme [88], and for screening for parameters for optimal displacement chromatography of proteins [78, 79]. Rational screening was also used for developing a displacement chromatography of proteoforms of a recombinant protein with HIC [89].

4.2 Separation of proteoforms of therapeutic proteins with capillary electrophoresis

Compared with liquid chromatography, capillary electrophoresis (CE) offers better resolving power. CE techniques such as capillary zone electrophoresis (CZE), capillary gel electrophoresis (CGE) and capillary isoelectric focusing (CIEF) have been adapted for the separation and characterization of proteins [90, 91]. These are basic techniques routinely used for quality control [91]. With CGE, the size of proteins is characterized, while in CIEF, proteins are separated according to their isoelectric point (pI). CIEF is using pH gradients formed by carrier ampholytes in a capillary [92]. It is important to note that pH plays a major role in CZE and should be well maintained [93]. Considerable protein adsorption must be considered when performing CIEF and CZE. The interaction of the analytes with the surface of the capillary may compromise the resolution, peak widths and shapes when using conventional bare fused-silica capillaries. Minimizing adsorption can be done by using better coating material or using reagents that reduce adsorption [94]. A penetrated surface layer protein A from bacteria was reported as capillary coating. The coating could be used for over 100 injections without loss of separation performance [95]. Another study reported that adsorption still happened when using LPA-coated capillary [96].

CZE and CIEF are more often used for separations of charge variants induced by C-terminal lysine truncation, N-terminal pyroglutamate formation, sialylation and deamidation [97].

The direct coupling of CE with MS is technically challenging regarding the CE-MS interface [98]. A study demonstrated a successful attempt to directly couple CIEF with mass spectrometry for characterization of transtuzumab, bevacizumab, cetuzimab and infliximab by optimizing the reagent, liquid composition and enhanced sample mixture by glycerol to reduce non-CIEF electrophoretic mobility and band broadening [99]. A CZE method was developed for the intact analysis of recombinant human interferon-β1 (rhIFN-β1). The charged species due to deamidation and sialylation were sufficiently separated. In contrast to dynamic polymeric coatings, such as polybrene or hydroxypropyl-methylcellulose, they covalently coated the bare-fused silica capillary with cross-linked polyethyleneimine (CPEI) to get positively charged surface, thus reducing the possibility of protein interaction with the coating. They then coupled this CZE to ESI-MS/MS and identified 138 proteoforms, of which, 55 were quantified.

For the in-depth characterization of the composition of proteoforms of a therapeutic protein CE online-coupled to MS is a good option, if prior to the CE, the mixture of proteoforms has already been fractionated by LC using separation mechanisms orthogonal to the CE separation mechanism.

5. Conclusion

A huge progress has been made in the field of TDMS, allowing the identification and comprehensive analysis of the composition of atoms of proteoforms, especially if they are smaller than 30 kDa. TDMS analysis of larger proteoforms still is more challenging. However, until today the most critical point is the purification of a proteoform towards near homogeneity or at least the significant reduction of complexity of the sample, which is desorbed and ionized into a tandem mass spectrometer for TDMS. A low complexity of the composition of a protein mixture entering the MS still is mandatory for getting high quality spectra. Thus, efficient separation methods are needed for obtaining fractions with low complexity. For developing strategies for separating proteoforms, therapeutic proteins are well suited, however challenging because of their heterogeneity. In depth separation of the proteoforms of a therapeutic protein requires the combination of fractionation techniques based on orthogonal mechanisms. In addition, the combination of gradient chromatography and displacement chromatography will add further opportunities for successful separations.

Acknowledgements

The authors would like to thank the European Research Council (ERC) for funding the A4B project by the Horizon 2020 Marie Sklodowska-Curie Action ITN 2017 of the European Commission (H2020-MSCA-ITN-2017; under the European Union's Horizon 2020 programme for research and innovation under grant agreement No 765502).

Conflict of interest

The authors declare no conflict of interest.

Abbreviations

AEX	anion exchange
ADCs	antibody-drug-conjugates
CE	capillary electrophoresis
CGE	capillary gel electrophoresis
CIEF	capillary isoelectric focusing
CZE	capillary zone electrophoresis
CEX	cation exchange
CID	collision-induced dissociation
ECD	electron capture dissociation
ETD	electron transfer dissociation
ETHcD	electron transfer higher energy collisional dissociation
ESI	electrospray ionization
GE	gradient elution
HCD	higher energy collisional dissociation
HAP	hydroxyapatite-chromatography
HILIC	hydrophilic interaction chromatography
HIC	hydrophobic interaction chromatography
IMAC	immobilized metal-affinity-chromatography

IRMPD	infrared multiple photon dissociation
IEX	ion exchange chromatography
LC	liquid chromatography
LC–MS/MS	liquid chromatography coupled to tandem mass spectrometry
MS	mass spectrometry
mAbs	monoclonal antibodies
PPS	protein-purification parameter screening system
rhIL	recombinant human interleukin
rTP	recombinant therapeutic protein
RPLC	reversed phase liquid chromatography
SEC	size exclusion chromatography
TDMS	top-down mass spectrometry
UVPD	ultraviolet photodissociation

Author details

Siti Nurul Hidayah[†], Manasi Gaikwad[†], Laura Heikaus and Hartmut Schlüter[*]
Mass Spectrometric Proteomics, Institute of Clinical Chemistry and Laboratory
Medicine, University Medical Center Hamburg-Eppendorf, Hamburg, Germany

*Address all correspondence to: hschluet@uke.de

† These authors have equally contributed as first authors.

IntechOpen

References

[1] Jungblut PR, Holzhutter HG, Apweiler R, Schluter H. The speciation of the proteome. Chemistry Central Journal. 2008;**2**:16

[2] Schluter H, Apweiler R, Holzhutter HG, Jungblut PR. Finding one's way in proteomics: A protein species nomenclature. Chemistry Central Journal. 2009;**3**:11

[3] Smith LM, Kelleher NL. Consortium for top down P. Proteoform: A single term describing protein complexity. Nature Methods. 2013;**10**(3):186-187

[4] Zhang Y, Xu T, Shan B, Hart J, Aslanian A, Han X, et al. ProteinInferencer: Confident protein identification and multiple experiment comparison for large scale proteomics projects. Journal of Proteomics. 2015;**129**:25-32

[5] Shliaha PV, Gibb S, Gorshkov V, Jespersen MS, Andersen GR, Bailey D, et al. Maximizing sequence coverage in top-down proteomics by automated multimodal gas-phase protein fragmentation. Analytical Chemistry. 2018;**90**(21):12519-12526

[6] Kou Q, Wu S, Liu X. Systematic evaluation of protein sequence filtering algorithms for proteoform identification using top-down mass spectrometry. Proteomics. 2018;**18**(3-4). PMID: 29327814

[7] Vyatkina K. De novo sequencing of top-down tandem mass spectra: A next step towards retrieving a complete protein sequence. Proteomes. 2017;**5**(1). PMID: 28248257

[8] Skinner OS, Havugimana PC, Haverland NA, Fornelli L, Early BP, Greer JB, et al. An informatic framework for decoding protein complexes by top-down mass spectrometry. Nature Methods. 2016;**13**(3):237-240

[9] Schaffer LV, Millikin RJ, Miller RM, Anderson LC, Fellers RT, Ge Y, et al. Identification and quantification of proteoforms by mass spectrometry. Proteomics. 2019;**19**(10):e1800361

[10] Donnelly DP, Rawlins CM, DeHart CJ, Fornelli L, Schachner LF, Lin Z, et al. Best practices and benchmarks for intact protein analysis for top-down mass spectrometry. Nature Methods. 2019;**16**(7):587-594

[11] Kelleher NL. A cell-based approach to the human proteome project. Journal of the American Society for Mass Spectrometry. 2012;**23**(10):1617-1624. PMID: 22976808

[12] Walsh G. Biopharmaceutical benchmarks 2014. Nature Biotechnology. 2014;**32**(10):992-1000

[13] Walsh G. Biopharmaceutical benchmarks 2018. Nature Biotechnology. 2018;**36**(12):1136-1145

[14] Lagasse HA, Alexaki A, Simhadri VL, Katagiri NH, Jankowski W, Sauna ZE, et al. Recent advances in (therapeutic protein) drug development. F1000Res. 2017;**6**:113. PMID: 28232867

[15] Lee AC, Harris JL, Khanna KK, Hong JH. A comprehensive review on current advances in peptide drug development and design. The International Journal of Molecular Sciences. 2019;**20**(10). PMID: 31091705

[16] Kim N, Cho D, Kim H, Kim S, Cha YJ, Greulich H, et al. Colorectal adenocarcinoma-derived EGFR mutants are oncogenic and sensitive to EGFR-targeted monoclonal antibodies, cetuximab and panitumumab. In: International Journal of Cancer Journal International du Cancer. 2019. DOI: 10.1002/ijc.32499. [Epub ahead of print] PMID: 31290142

[17] Chen M, Manley JL. Mechanisms of alternative splicing regulation: Insights from molecular and genomics approaches. Nature Reviews. Molecular Cell Biology. 2009;**10**(11):741-754

[18] Sarmiento C, Camarero JA. Biotechnological applications of protein splicing. Current Protein and Peptide Science. 2019;**20**(5):408-424

[19] Mathur M, Kim CM, Munro SA, Rudina SS, Sawyer EM, Smolke CD. Programmable mutually exclusive alternative splicing for generating RNA and protein diversity. Nature Communications. 2019;**10**(1):2673

[20] Caval T, Tian W, Yang Z, Clausen H, Heck AJR. Direct quality control of glycoengineered erythropoietin variants. Nature Communications. 2018;**9**(1):3342

[21] Fazel R, Guan Y, Vaziri B, Krisp C, Heikaus L, Saadati A, et al. Structural and in vitro functional comparability analysis of Altebrel, a proposed Etanercept biosimilar: Focus on primary sequence and glycosylation. Pharmaceuticals (Basel). 2019;**12**(1). PMID: 30658444

[22] Hansen HG, Pristovsek N, Kildegaard HF, Lee GM. Improving the secretory capacity of Chinese hamster ovary cells by ectopic expression of effector genes: Lessons learned and future directions. Biotechnology Advances. 2017;**35**(1):64-76

[23] Le Fourn V, Girod PA, Buceta M, Regamey A, Mermod N. CHO cell engineering to prevent polypeptide aggregation and improve therapeutic protein secretion. Metabolic Engineering. 2014;**21**:91-102

[24] Estes B, Hsu YR, Tam LT, Sheng J, Stevens J, Haldankar R. Uncovering methods for the prevention of protein aggregation and improvement of

product quality in a transient expression system. Biotechnology Progress. 2015;**31**(1):258-267

[25] Johari YB, Estes SD, Alves CS, Sinacore MS, James DC. Integrated cell and process engineering for improved transient production of a "difficult-to-express" fusion protein by CHO cells. Biotechnology and Bioengineering. 2015;**112**(12):2527-2542

[26] Gikanga B, Eisner DR, Ovadia R, Day ES, Stauch OB, Maa YF. Processing impact on monoclonal antibody drug products: Protein subvisible particulate formation induced by grinding stress. PDA Journal of Pharmaceutical Science and Technology. 2017;**71**(3):172-188

[27] Rathore N, Rajan RS. Current perspectives on stability of protein drug products during formulation, fill and finish operations. Biotechnology Progress. 2008;**24**(3):504-514

[28] Seidl A, Hainzl O, Richter M, Fischer R, Bohm S, Deutel B, et al. Tungsten-induced denaturation and aggregation of epoetin alfa during primary packaging as a cause of immunogenicity. Pharmaceutical Research. 2012;**29**(6):1454-1467

[29] Jefferis R. Glycosylation as a strategy to improve antibody-based therapeutics. Nature Reviews. Drug Discovery. 2009;**8**(3):226-234

[30] Lingg N, Zhang P, Song Z, Bardor M. The sweet tooth of biopharmaceuticals: Importance of recombinant protein glycosylation analysis. Biotechnology Journal. 2012;**7**(12):1462-1472

[31] D'Souza AJ, Mar KD, Huang J, Majumdar S, Ford BM, Dyas B, et al. Rapid deamidation of recombinant protective antigen when adsorbed on aluminum hydroxide gel correlates with reduced potency of vaccine.

Journal of Pharmaceutical Sciences. 2013;**102**(2):454-461

[32] Nellis DF, Michiel DF, Jiang MS, Esposito D, Davis R, Jiang H, et al. Characterization of recombinant human IL-15 deamidation and its practical elimination through substitution of asparagine 77. Pharmaceutical Research. 2012;**29**(3):722-738

[33] Ke N, Berkmen M. Production of disulfide-bonded proteins in *Escherichia coli*. Current Protocols in Molecular Biology. 2014;**108**:16 1B 1-16 1B21

[34] Du C, Huang Y, Borwankar A, Tan Z, Cura A, Yee JC, et al. Using hydrogen peroxide to prevent antibody disulfide bond reduction during manufacturing process. MAbs. 2018;**10**(3):500-510

[35] Lakbub JC, Shipman JT, Desaire H. Recent mass spectrometry-based techniques and considerations for disulfide bond characterization in proteins. Analytical and Bioanalytical Chemistry. 2018;**410**(10):2467-2484

[36] Beyer B, Schuster M, Jungbauer A, Lingg N. Microheterogeneity of recombinant antibodies: Analytics and functional impact. Biotechnology Journal. 2018;**13**(1)

[37] Ambrogelly A, Gozo S, Katiyar A, Dellatore S, Kune Y, Bhat R, et al. Analytical comparability study of recombinant monoclonal antibody therapeutics. MAbs. 2018;**10**(4):513-538

[38] Reichel C, Thevis M. Gel electrophoretic methods for the analysis of biosimilar pharmaceuticals using the example of recombinant erythropoietin. Bioanalysis. 2013;**5**(5):587-602

[39] Haverick M, Mengisen S, Shameem M, Ambrogelly A. Separation of mAbs molecular variants by analytical hydrophobic interaction chromatography HPLC: Overview and applications. MAbs. 2014;**6**(4):852-858

[40] Roberts CJ. Therapeutic protein aggregation: Mechanisms, design, and control. Trends in Biotechnology. 2014;**32**(7):372-380

[41] Arbogast LW, Brinson RG, Marino JP. Mapping monoclonal antibody structure by 2D 13C NMR at natural abundance. Analytical Chemistry. 2015;**87**(7):3556-3561

[42] Zhou Y, Raju R, Alves C, Gilbert A. Debottlenecking protein secretion and reducing protein aggregation in the cellular host. Current Opinion in Biotechnology. 2018;**53**:151-157

[43] Moussa EM, Panchal JP, Moorthy BS, Blum JS, Joubert MK, Narhi LO, et al. Immunogenicity of therapeutic protein aggregates. Journal of Pharmaceutical Sciences. 2016;**105**(2):417-430

[44] Sorensen M, Harmes DC, Stoll DR, Staples GO, Fekete S, Guillarme D, et al. Comparison of originator and biosimilar therapeutic monoclonal antibodies using comprehensive two-dimensional liquid chromatography coupled with time-of-flight mass spectrometry. MAbs. 2016;**8**(7):1224-1234

[45] Ramos-de-la-Pena AM, Gonzalez-Valdez J, Aguilar O. Protein a chromatography: Challenges and progress in the purification of monoclonal antibodies. Journal of Separation Science. 2019;**42**(9):1816-1827

[46] Khawli LA, Goswami S, Hutchinson R, Kwong ZW, Yang J, Wang X, et al. Charge variants in IgG1: Isolation, characterization, in vitro binding properties and pharmacokinetics in rats. MAbs. 2010;**2**(6):613-624

[47] Hintersteiner B, Lingg N, Zhang P, Woen S, Hoi KM, Stranner S, et al. Charge heterogeneity: Basic antibody charge variants with increased binding to fc receptors. MAbs. 2016;**8**(8):1548-1560

[48] Fussl F, Cook K, Scheffler K, Farrell A, Mittermayr S, Bones J. Charge variant analysis of monoclonal antibodies using direct coupled pH gradient cation exchange chromatography to high-resolution native mass spectrometry. Analytical Chemistry. 2018;**90**(7):4669-4676

[49] Du Y, Walsh A, Ehrick R, Xu W, May K, Liu H. Chromatographic analysis of the acidic and basic species of recombinant monoclonal antibodies. MAbs. 2012;**4**(5):578-585

[50] Sankaran PK, Kabadi PG, Honnappa CG, Subbarao M, Pai HV, Adhikary L, et al. Identification and quantification of product-related quality attributes in bio-therapeutic monoclonal antibody via a simple, and robust cation-exchange HPLC method compatible with direct online detection of UV and native ESI-QTOF-MS analysis. Journal of chromatography B. Analytical technologies in the biomedical and life sciences. 2018;**1102-1103**:83-95

[51] Zhang Q, Yang FQ, Ge L, Hu YJ, Xia ZN. Recent applications of hydrophilic interaction liquid chromatography in pharmaceutical analysis. Journal of Separation Science. 2017;**40**(1):49-80

[52] Ikegami T. Hydrophilic interaction chromatography for the analysis of biopharmaceutical drugs and therapeutic peptides: A review based on the separation characteristics of the hydrophilic interaction chromatography phases. Journal of Separation Science. 2019;**42**(1):130-213

[53] Chen B, Lin Z, Alpert AJ, Fu C, Zhang Q, Pritts WA, et al. Online hydrophobic interaction chromatography-mass spectrometry for the analysis of intact monoclonal antibodies. Analytical Chemistry. 2018;**90**(12):7135-7138

[54] Boyd D, Kaschak T, Yan B. HIC resolution of an IgG1 with an oxidized Trp in a complementarity determining region. Journal of Chromatography B, Analytical Technologies in the Biomedical and Life Sciences. 2011;**879**(13-14):955-960

[55] Bobaly B, Randazzo GM, Rudaz S, Guillarme D, Fekete S. Optimization of non-linear gradient in hydrophobic interaction chromatography for the analytical characterization of antibody-drug conjugates. Journal of Chromatography. A. 2017;**1481**:82-91

[56] Valliere-Douglass J, Wallace A, Balland A. Separation of populations of antibody variants by fine tuning of hydrophobic-interaction chromatography operating conditions. Journal of Chromatography. A. 2008;**1214**(1-2):81-89

[57] Ouellette D, Chumsae C, Clabbers A, Radziejewski C, Correia I. Comparison of the in vitro and in vivo stability of a succinimide intermediate observed on a therapeutic IgG1 molecule. MAbs. 2013;**5**(3):432-444

[58] Harris RJ. Heterogeneity of recombinant antibodies: Linking structure to function. Developmental Biology (Basel). 2005;**122**:117-127

[59] Halan V, Maity S, Bhambure R, Rathore AS. Multimodal chromatography for purification of biotherapeutics—A review. Current Protein and Peptide Science. 2019;**20**(1):4-13

[60] Schluter H, Zidek W. Application of non-size-related separation effects

to the purification of biologically-active substances with a size-exclusion gel. Journal of Chromatography. 1993;**639**(1):17-22

[61] Hong P, Koza S, Bouvier ES. Size-exclusion chromatography for the analysis of protein biotherapeutics and their aggregates. Journal of Liquid Chromatography and Related Technologies. 2012;**35**(20):2923-2950

[62] Zolls S, Tantipolphan R, Wiggenhorn M, Winter G, Jiskoot W, Friess W, et al. Particles in therapeutic protein formulations, part 1: Overview of analytical methods. Journal of Pharmaceutical Sciences. 2012;**101**(3):914-935

[63] Turner A, Yandrofski K, Telikepalli S, King J, Heckert A, Filliben J, et al. Development of orthogonal NISTmAb size heterogeneity control methods. Analytical and Bioanalytical Chemistry. 2018;**410**(8):2095-2110

[64] Mohamed HE, Mohamed AA, Al-Ghobashy MA, Fathalla FA, Abbas SS. Stability assessment of antibody-drug conjugate Trastuzumab emtansine in comparison to parent monoclonal antibody using orthogonal testing protocol. Journal of Pharmaceutical and Biomedical Analysis. 2018;**150**:268-277

[65] Yang Y, Li H, Li Z, Zhang Y, Zhang S, Chen Y, et al. Size-exclusion HPLC provides a simple, rapid, and versatile alternative method for quality control of vaccines by characterizing the assembly of antigens. Vaccine. 2015;**33**(9):1143-1150

[66] Ehkirch A, D'Atri V, Rouviere F, Hernandez-Alba O, Goyon A, Colas O, et al. An online four-dimensional HICxSEC-IMxMS methodology for proof-of-concept characterization of antibody drug conjugates. Analytical Chemistry. 2018;**90**(3):1578-1586

[67] Wang X, Chen L. Challenges in bioanalytical assays for biosimilars. Bioanalysis. 2014;**6**(16):2111-2113

[68] Le JC, Bondarenko PV. Trap for MAbs: Characterization of intact monoclonal antibodies using reversed-phase HPLC on-line with ion-trap mass spectrometry. Journal of the American Society for Mass Spectrometry. 2005;**16**(3):307-311

[69] Ambrogelly A, Liu YH, Li H, Mengisen S, Yao B, Xu W, et al. Characterization of antibody variants during process development: The tale of incomplete processing of N-terminal secretion peptide. MAbs. 2012;**4**(6):701-709

[70] Dillon TM, Bondarenko PV, Rehder DS, Pipes GD, Kleemann GR, Ricci MS. Optimization of a reversed-phase high-performance liquid chromatography/mass spectrometry method for characterizing recombinant antibody heterogeneity and stability. Journal of Chromatography. A. 2006;**1120**(1-2):112-120

[71] Bobaly B, D'Atri V, Lauber M, Beck A, Guillarme D, Fekete S. Characterizing various monoclonal antibodies with milder reversed phase chromatography conditions. Journal of chromatography B, analytical technologies in the biomedical and Life Sciences. 2018;**1096**:1-10

[72] Jankowski J, Potthoff W, Zidek W, Schluter H. Purification of chemically synthesised dinucleoside(5′,5′) polyphosphates by displacement chromatography. Journal of Chromatography. B, Biomedical Sciences and Applications. 1998;**719**(1-2):63-70

[73] Trusch M, Bohlick A, Hildebrand D, Lichtner B, Bertsch A, Kohlbacher O, et al. Application of displacement chromatography for the analysis of a lipid

raft proteome. Journal of Chromatography B, Analytical Technologies in the Biomedical and Life Sciences. 2010;**878**(3-4):309-314

[74] Trusch M, Tillack K, Kwiatkowski M, Bertsch A, Ahrends R, Kohlbacher O, et al. Displacement chromatography as first separating step in online two-dimensional liquid chromatography coupled to mass spectrometry analysis of a complex protein sample—The proteome of neutrophils. Journal of Chromatography. A. 2012;**1232**:288-294

[75] Kwiatkowski M, Krosser D, Wurlitzer M, Steffen P, Barcaru A, Krisp C, et al. Application of displacement chromatography to online two-dimensional liquid chromatography coupled to tandem mass spectrometry improves peptide separation efficiency and detectability for the analysis of complex proteomes. Analytical Chemistry. 2018;**90**(16):9951-9958

[76] Ahrends R, Lichtner B, Bertsch A, Kohlbacher O, Hildebrand D, Trusch M, et al. Application of displacement chromatography for the proteome analysis of a human plasma protein fraction. Journal of Chromatography. A. 2010;**1217**(19):3321-3329

[77] Ahrends R, Lichtner B, Buck F, Hildebrand D, Kotasinska M, Kohlbacher O, et al. Comparison of displacement versus gradient mode for separation of a complex protein mixture by anion-exchange chromatography. Journal of Chromatography B, Analytical Technologies in the Biomedical and Life Sciences. 2012;**901**:34-40

[78] Kotasinska M, Richter V, Kwiatkowski M, Schluter H. Sample displacement batch chromatography of proteins. Methods in Molecular Biology. 2014;**1129**:325-338

[79] Kotasinska M, Richter V, Thiemann J, Schluter H. Cation exchange displacement batch chromatography of proteins guided by screening of protein purification parameters. Journal of Separation Science. 2012;**35**(22):3170-3176

[80] Khanal O, Kumar V, Westerberg K, Schlegel F, Lenhoff AM. Multi-column displacement chromatography for separation of charge variants of monoclonal antibodies. Journal of Chromatography. A. 2019;**1586**:40-51

[81] Ren J, Yao P, Chen J, Jia L. Salt-independent hydrophobic displacement chromatography for antibody purification using cyclodextrin as supermolecular displacer. Journal of Chromatography. A. 2014;**1369**:98-104

[82] McAtee CP, Hornbuckle J. Isolation of monoclonal antibody charge variants by displacement chromatography. Current Protocols in Protein Science. 2012;**8**:10

[83] Zhang T, Bourret J, Cano T. Isolation and characterization of therapeutic antibody charge variants using cation exchange displacement chromatography. Journal of Chromatography. A. 2011;**1218**(31):5079-5086

[84] Luellau E, von Stockar U, Vogt S, Freitag R. Development of a downstream process for the isolation and separation of monoclonal immunoglobulin a monomers, dimers and polymers from cell culture supernatant. Journal of Chromatography. A. 1998;**796**(1):165-175

[85] Mazza CB, Rege K, Breneman CM, Sukumar N, Dordick JS, Cramer SM. High-throughput screening and quantitative structure-efficacy relationship models of potential displacer molecules for ion-exchange systems. Biotechnology and Bioengineering. 2002;**80**(1):60-72

[86] Tugcu N, Ladiwala A, Breneman CM, Cramer SM. Identification of chemically selective displacers using parallel batch screening experiments and quantitative structure efficacy relationship models. Analytical Chemistry. 2003;**75**(21):5806-5816

[87] Thiemann J, Jankowski J, Rykl J, Kurzawski S, Pohl T, Wittmann-Liebold B, et al. Principle and applications of the protein-purification-parameter screening system. Journal of Chromatography. A. 2004;**1043**(1):73-80

[88] Rykl J, Thiemann J, Kurzawski S, Pohl T, Gobom J, Zidek W, et al. Renal cathepsin G and angiotensin II generation. Journal of Hypertension. 2006;**24**(9):1797-1807

[89] Sunasara KM, Xia F, Gronke RS, Cramer SM. Application of hydrophobic interaction displacement chromatography for an industrial protein purification. Biotechnology and Bioengineering. 2003;**82**(3):330-339

[90] Voeten RLC, Ventouri IK, Haselberg R, Somsen GW. Capillary electrophoresis: Trends and recent advances. Analytical Chemistry. 2018;**90**(3):1464-1481

[91] Suba D, Urbanyi Z, Salgo A. Capillary isoelectric focusing method development and validation for investigation of recombinant therapeutic monoclonal antibody. Journal of Pharmaceutical and Biomedical Analysis. 2015;**114**:53-61

[92] Koshel BM, Wirth MJ. Trajectory of isoelectric focusing from gels to capillaries to immobilized gradients in capillaries. Proteomics. 2012;**12**(19-20):2918-2926

[93] Zhu G, Sun L, Dovichi NJ. Dynamic pH junction preconcentration in capillary electrophoresis- electrospray ionization-mass spectrometry for proteomics analysis. The Analyst. 2016;**141**(18):5216-5220

[94] de Jong S, Epelbaum N, Liyanage R, Krylov SN. A semipermanent coating for preventing protein adsorption at physiological pH in kinetic capillary electrophoresis. Electrophoresis. 2012;**33**(16):2584-2590

[95] Yu B, Peng Q, Usman M, Ahmed A, Chen Y, Chen X, et al. Preparation of photosensitive diazotized poly (vinyl alcohol-b-styrene) covalent capillary coatings for capillary electrophoresis separation of proteins. Journal of Chromatography. A. 2019;**1593**:174-182

[96] Graf M, Watzig H. Capillary isoelectric focusing—Reproducibility and protein adsorption. Electrophoresis. 2004;**25**(17):2959-2964

[97] Gahoual R, Beck A, Leize-Wagner E, Francois YN. Cutting-edge capillary electrophoresis characterization of monoclonal antibodies and related products. Journal of Chromatography B, Analytical Technologies in the Biomedical and Life Sciences. 2016;**1032**:61-78

[98] Huhner J, Jooss K, Neususs C. Interference-free mass spectrometric detection of capillary isoelectric focused proteins, including charge variants of a model monoclonal antibody. Electrophoresis. 2017;**38**(6):914-921

[99] Dai J, Lamp J, Xia Q, Zhang Y. Capillary isoelectric focusing-mass spectrometry method for the separation and online characterization of intact monoclonal antibody charge variants. Analytical Chemistry. 2018;**90**(3):2246-2254

[100] Arakawa T, Tsumoto K, Ejima D. Alternative downstream processes for production of antibodies and antibody fragments. Biochimica et Biophysica Acta. 2014;**1844**(11):2032-2040

[101] Yigzaw Y, Hinckley P, Hewig A, Vedantham G. Ion exchange chromatography of proteins and clearance of aggregates. Current Pharmaceutical Biotechnology. 2009;**10**(4):421-426

[102] Fekete S, Beck A, Veuthey JL, Guillarme D. Ion-exchange chromatography for the characterization of biopharmaceuticals. Journal of Pharmaceutical and Biomedical Analysis. 2015;**113**:43-55

[103] Ponniah G, Kita A, Nowak C, Neill A, Kori Y, Rajendran S, et al. Characterization of the acidic species of a monoclonal antibody using weak cation exchange chromatography and LC-MS. Analytical Chemistry. 2015;**87**(17):9084-9092

[104] Cummings LJ, Snyder MA, Brisack K. Protein chromatography on hydroxyapatite columns. Methods in Enzymology. 2009;**463**:387-404

[105] Block H, Maertens B, Spriestersbach A, Brinker N, Kubicek J, Fabis R, et al. Immobilized-metal affinity chromatography (IMAC): A review. Methods in Enzymology. 2009;**463**:439-473

[106] Cheung RC, Wong JH, Ng TB. Immobilized metal ion affinity chromatography: A review on its applications. Applied Microbiology and Biotechnology. 2012;**96**(6):1411-1420

[107] Fekete S, Beck A, Wagner E, Vuignier K, Guillarme D. Adsorption and recovery issues of recombinant monoclonal antibodies in reversed-phase liquid chromatography. Journal of Separation Science. 2015;**38**(1):1-8

[108] Goyon A, Fekete S, Beck A, Veuthey JL, Guillarme D. Unraveling the mysteries of modern size exclusion chromatography—The way to achieve confident characterization of therapeutic proteins. Journal

of chromatography B, analytical technologies in the biomedical and. Life Sciences. 2018;**1092**:368-378

[109] Rathore AS, Kumar D, Kateja N. Recent developments in chromatographic purification of biopharmaceuticals. Biotechnology Letters. 2018;**40**(6):895-905

[110] Wagh A, Song H, Zeng M, Tao L, Das TK. Challenges and new frontiers in analytical characterization of antibody-drug conjugates. MAbs. 2018;**10**(2):222-243

[111] Stoll D, Danforth J, Zhang K, Beck A. Characterization of therapeutic antibodies and related products by two-dimensional liquid chromatography coupled with UV absorbance and mass spectrometric detection. Journal of Chromatography B, Analytical Technologies in the Biomedical and Life Sciences. 2016;**1032**:51-60

[112] Schlecht J, Jooss K, Neususs C. Two-dimensional capillary electrophoresis-mass spectrometry (CE-CE-MS): Coupling MS-interfering capillary electromigration methods with mass spectrometry. Analytical and Bioanalytical Chemistry. 2018;**410**(25):6353-6359

[113] Nebija D, Noe CR, Urban E, Lachmann B. Quality control and stability studies with the monoclonal antibody, trastuzumab: Application of 1D- vs. 2D-gel electrophoresis. International Journal of Molecular Sciences. 2014;**15**(4):6399-6411

Chapter 4

Prolactin Proteoform Pattern Changed in Human Pituitary Adenoma Relative to Control Pituitary Tissues

Xianquan Zhan and Shehua Qian

Abstract

PRL gene-encoded prolactin is synthesized in the ribosome in the pituitary and then secretes into blood circulation to reach its target organ and exerts its biological roles, for example, involving in production, growth, development, immunoregulation, and metabolism. Multiple post-translational modifications and other unknown factors might be involved in this process to cause different prolactin proteoforms with differential isoelectric point (pI) and relative mass (M_r). Pituitary adenomas are the common disease occurring in pituitary organ to affect the endocrine system. Two-dimensional gel electrophoresis (2DGE) was used to separate prolactin proteoforms according to their pI and M_r, followed by identification with Western blot and mass spectrometry (MS) analyses. Six prolactin proteoforms were identified in control pituitary tissues, and this prolactin proteoform pattern was significantly changed in different hormone subtypes of nonfunctional pituitary adenomas (NF$^-$, LH$^+$, FSH$^+$, and LH$^+$/FSH$^+$) and prolactinomas (PRL$^+$). Further, bioinformatics analysis revealed that different prolactin proteoforms might bind to different short- or long-PRL receptor-mediated signaling pathways. These findings clearly demonstrated that prolactin proteoform pattern existed in human pituitary and changed in different subtypes of pituitary adenomas. It is the scientific data to in-depth study prolactin functions, and to discover the prolactin proteoform biomarkers for PRL-related adenomas.

Keywords: prolactin proteoforms, pituitary, pituitary adenomas, nonfunctional pituitary adenomas, prolactinomas, biomarker

1. Introduction

Prolactin (PRL) is a multifunctional hormone which is synthesized and secreted by pituitary [1]. Human PRL gene is located on chromosome 6 [2]. The secretory mode of PRL is autocrine and paracrine [3], and the secretion of PRL is pulsating and circadian rhythm [4]. The concentration of PRL in human serum has a certain reference range, and when its concentration is too high or too low, it will have a certain impact on the body. Dopamine can inhibit the secretion of

PRL, and there are cases where dopamine is used to treat hyperprolactinemia [1]. PRL's biological functions include production, growth, development, immunoregulation, and metabolism [5, 6]. PRL can exert its biological functions only when it binds to its receptor and activates some signaling pathways [7]. According to the concept of proteoforms, a protein is defined as a set of proteoforms, due to different splicing, post-translational modifications (PTMs), and even unknown factors. Each proteoform has its own specific isoelectric point (pI) and molecular weight (M_r). For human PRL in the UniProt protein database, its pI is 6.5 and M_r is 25.88 kDa. However, Ben-Jonathan et al. found that human serum contained PRLs with M_r > 100, 40–60 and 16 kDa, besides the PRL with 25.88 kDa [8]. Qian et al. found six PRLs with different pI and M_r in human pituitary tissues by two-dimensional gel electrophoresis (2DGE) and mass spectrometry (MS) [9]. Similarly, Zhan et al. found 24 growth hormone (GH) with different pI and M_r in human pituitary tissues by 2-DGE and MS [10]. A possible reason of this difference of pI and M_r in human PRL and GH is that they undergo PTMs or splicing [11]. A proteoform is a specific form that protein exerts its final functions, which is derived from a gene undergoing splicing, transcription, translation, PTMs, translocation/re-distribution, and interaction with other molecules, etc. [12].

Recently 2DGE and MS have been recognized as high throughput and useful tool to study proteoforms [13–15]. 2DGE is able to separate each proteoform in the first dimension—isoelectric focusing (IEF) based on proteoform charge difference, and in the second dimension—sodium dodecyl sulfate polyacrylamide gel electrophoresis (SDS-PAGE) based on proteoform relative mass difference [16]. Therefore, 2DGE achieves proteoform separation based on the difference in pI and M_r of proteoforms. And then the protein (exactly proteoforms) on the 2D gel is transferred to a polyvinylidene fluoride membrane for detection with a specific antibody. The immunoreactive positive 2D gel spots represent the proteoforms of a protein. The proteoform in each immunoreactive positive spot was subjected to in-gel digestion with trypsin, and identification with MS [17, 18]. MS is a key technique to identify organic molecules and analyze the extreme structure of certain substances [19]. Especially, top-down MS can quickly and extremely accurately determine the molecular weight of biomacromolecules, which enables proteomics research from protein general identification to advanced structural studies and protein-protein interaction studies. Moreover, with the development of MS technology, the accuracy and sensitivity of mass spectrometers have been greatly improved. MS has its absolute advantages in the use of less sample, faster analysis, and simultaneous separation and identification. Therefore, 2DGE in combination with MS is presenting as a super-high approach in separation and identification of large-scale human proteoforms [14]. If the stable isotope labeling is introduced to prepare the protein sample prior to 2DGE-MS, then 2DGE-MS can also quantify the abundance of a proteoform between two given conditions such as tumors vs. controls [20].

Pituitary adenomas are the common disease occurred in pituitary organ to severely impact on the human endocrine system. PRL is an important pituitary hormone. It has important scientific merit in clarification of PRL proteoform pattern changed in different subtypes of pituitary adenomas compared to control pituitaries. This book chapter focuses on the PRL proteoforms in human pituitary and investigates the PRL proteoform pattern alterations in pituitary adenomas relative to controls, with 2DGE and MS. These findings provide the scientific data to in-depth study the PRL functions and to discover PRL proteoform biomarker for PRL-related adenomas.

2. Methods

2.1 Pituitary tissue samples and preparation of protein samples

Eight human post-mortem control pituitary tissues, five PRL-positive prolactinoma tissues, three non-hormone expressed nonfunctional pituitary adenoma (NF-NFPA) tissues, three luteinizing hormone (LH)-positive NFPA tissues, three follicle-stimulating hormone (FSH)-positive NFPA tissues, and three LH/FSH-both positive NFPA tissues were used to extract proteins, with the previously described procedure [21, 22]. The extracted protein of each tissue sample was used for 2DGE and MS analysis.

2.2 2DGE

A amount (70 μg) of proteins was diluted into 350 μL of protein sample buffer (7 mol/L urea, 2 mol/L thiourea, 40 g/L CHAPS, 100 mmol/L dithiothreitol (DTT), 5 mol/L immobilized pH gradient (IPG) buffer pH 3–10 NL, and a trace of bromophenol blue, followed by rehydration (18 h, 20°C) of precast IPG strips pH 3–10 NL (180 x 3 × 0.5 mm) in 18-cm IPG strip holder on an IPGphor instrument, and IEF (25°C) with parameters (Gradient 250 V and 1 h for 125 Vh, gradient at 1000 V and 1 h for 500 Vh, gradient at 8000 V and 1 h for 4000 Vh, step-and-hold at 8000 and 4 h for 32,000 Vh, step-and-hold at 500 V and 0.5 h for 250 Vh to achieve a total of 36,875 Vh). After IEF, the proteins on IPG strip were reduced (15 min) with DTT, and alkylated with iodoacetamide, followed by separation with 12% SDS-PAGE (250 × 215 × 1.0 mm) in an Ettan DALT II system (GE Healthcare, up to 12 gels at a time) with a constant voltage (250 V, 360 min). All 2DGE-arrayed proteins were stained with silver-staining [23], and then digitized and analyzed with Discovery Series PDQuest 2D Gel Analysis software [24, 25]. Each sample was performed for 3–5 times.

2.3 2DGE-based Western blot

The proteins in the 2D gel were partially transferred to a polyvinylidene fluoride (PVDF) membrane (0.8 mA/cm^2 for 80 min) using Amersham Pharmacia Biotech Nova Blot semi-dry transfer instrument, followed by blocking (1 h, room temperature) with bovine serum albumin (BSA), incubation (1 h, room temperature) with rabbit anti-hPRL antibodies, incubation (1 h, room temperature) with goat anti-rabbit alkaline phosphatase conjugated IgG, and visualization with 5-bromo-4-chloro-3-indolyl phosphate. The detailed procedure was described previously [9].

2.4 In-gel digestion with trypsin and MS identification of PRL

The proteins in each Western blot-positive spot was performed in-gel digestion with trypsin, purification of tryptic peptides with ZipTipC18, followed by analysis with three types of MS instruments, including MALDI-TOF MS [24], LC-ESI-Q-IT MS [24], and MALDI-TOF-TOF MS [9]. The detailed procedure was described previously [9, 24]. The obtained peptide mass fingerprint (PMF) and tandem mass spectrometry (MS/MS) data were used to search Swiss-Prot human database for protein determination and PTM analysis.

2.5 Bioinformatics and statistical analysis

The phosphorylation sites, O-glycosylation sites, and N-glycosylation sites in the PRL amino acid sequence were predicted with NetPhos 3.1 Server (http://www.cbs.dtu.dk/services/NetPhos) [26, 27], NetOGlyc 4.0 Server (http://www.cbs.dtu.dk/services/NetOGlyc) [28], and NetNGlyc 1.0 Server (http://www.cbs.dtu.dk/services/NetNGlyc) [29]. The PRL proteoform pattern changes were tested with the Chi-square test among different subtypes of pituitary adenomas (p < 0.05).

3. Results and discussion

3.1 The amino acid sequences of human PRL prohormone and mature PRL

In human pituitary, the PRL prohormone is synthesized in the ribosome, with 227 amino acids (position 1–227; 25.9 kDa), containing a signal peptide (position 1–28) (**Table 1**), which was assigned with Swiss-Prot accession No. P01236. However, the mature human PRL only contains 199 amino acids (position 29–227; 22.9 kDa), which removed the signal peptide (position 1–28), and secreted into the circulation system to bind to its target organ for exerting PRL function.

3.2 PRL proteoform pattern in human pituitaries

The PRL proteoform pattern was found in human pituitaries. Qian et al. [9] found six PRL proteoforms with 2DGE in human pituitaries and then verified four of six PRL proteoforms with 2DGE-based Western blot in human pituitaries (**Figures 1** and **2**). The pI and M_r of these PRL proteoforms are slightly different. Each PRL proteoform was digested with trypsin, and followed by MS and MS/MS analysis (**Figures 3** and **4**). The characteristic tryptic peptide are calculated to determine whether the signal peptide (position 1–28) in each PRL proteoform (**Table 2**), which was compared to the observed ions of each PRL proteoform. It found all PRL proteoforms all contained the tryptic peptide sequence MNIKGSPWK (position 1–9), which clearly demonstrated that six PRL proteoforms are all PRL prohormone, but not mature PRL.

3.3 PRL proteoform changes in human pituitary adenomas compared to controls

The PRL proteoform pattern changed in different subtypes of pituitary adenomas compared to control pituitaries (**Table 3**). The ratio of each subtype of pituitary

10	20	30	40	50
MNIKGSPWKG	**SLLLLLVSNL**	**LLCQSVAP**LP	ICPGGAARCQ	VTLRDLFDRA
60	70	80	90	100
VVLSHYIHNL	SSEMFSEFDK	RYTHGRGFIT	KAINSCHTSS	LATPEDKEQA
110	120	130	140	150
QQMNQKDFLS	LIVSILRSWN	EPLYHLVTEV	RGMQEAPEAI	LSKAVEIEEQ
160	170	180	190	200
TKRLLEGMEL	IVSQVHPETK	ENEIYPVWSG	LPSLQMADEE	SRLSAYYNLL
210	220			
HCLRRDSHKI	DNYLKLLKCR	IIHNNNC		

Reproduced from Qian et al. [9], with copyright permission from Frontiers in open access article, copyright 2018.

Table 1.
Human PRL prohormone amino acid sequence (position 1–227) and mature PRL (position 29–227). The signal peptide is position 1-28 in the bold letters.

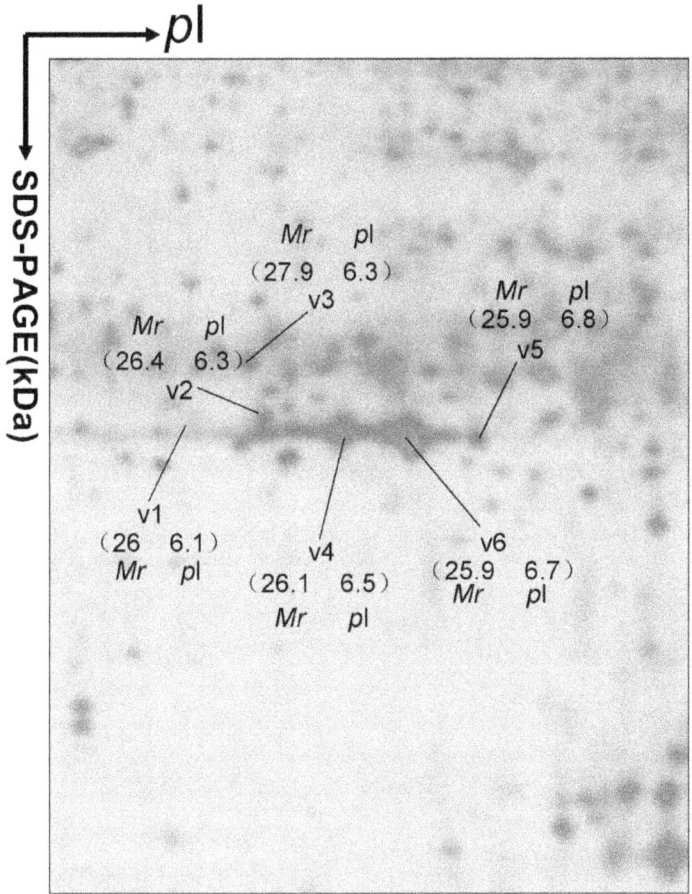

Figure 1.
PRL proteoform pattern in human pituitaries with a 2DGE gel image. Reproduced from Qian et al. [9], with copyright permission from Frontiers in open access article, copyright 2018.

adenoma relative to control pituitaries was decreased or unchanged. The proportional ratio of six PRL proteoforms among five subtypes of pituitary adenomas was changed (**Table 4** and **Figure 5**). In FSH⁺/LH⁺ and PRL⁺ pituitary adenomas, the proportion of PRL proteoform v1 is the largest. In FSH⁺ pituitary adenoma, the proportion of PRL proteoform v5 is the largest. The PRL proteoform changes suggest their scientific merit for clinical application.

3.4 Bioinformatics prediction of potential factors to form PRL proteoforms and pathway networks

PRL is a hormone which is secreted by pituitary gland. PRL has a variety of biological functions. Only when it reaches a specific target organ and binds to its receptor can it play its biological function (**Figure 6**). PRL can bind to short PRL receptor or long PRL receptor and then plays its biological functions. The long or short PRL receptors definitely bind to different PRL proteoforms. PRL proteoforms are definitely derived from a PRL gene undergoing splicing, transcription, translation, PTMs, translocation/re-distribution, and interaction with other molecules, etc. Therefore, phosphorylation sites in hPRL (position 1–227) were predicted

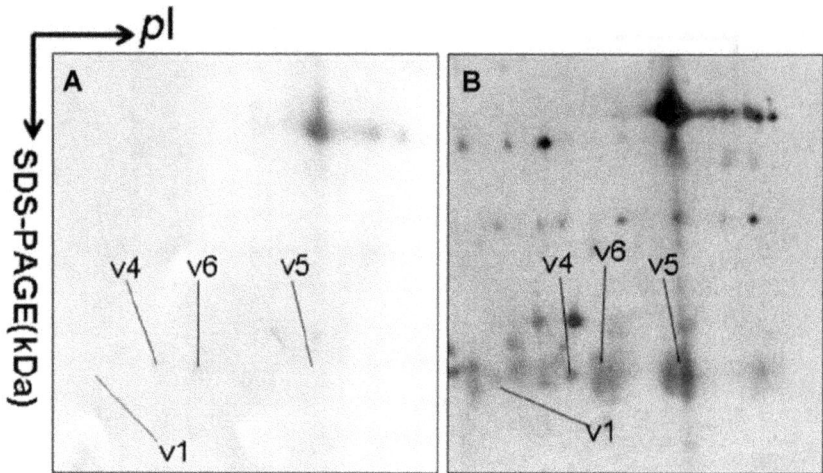

Figure 2.
Verification of PRL proteoforms with 2DGE-based Western blot in human pituitaries. (A) Western blot image. (B) Silver-stained image. Reproduced from Qian et al. [9], with copyright permission from Frontiers in open access article, copyright 2018.

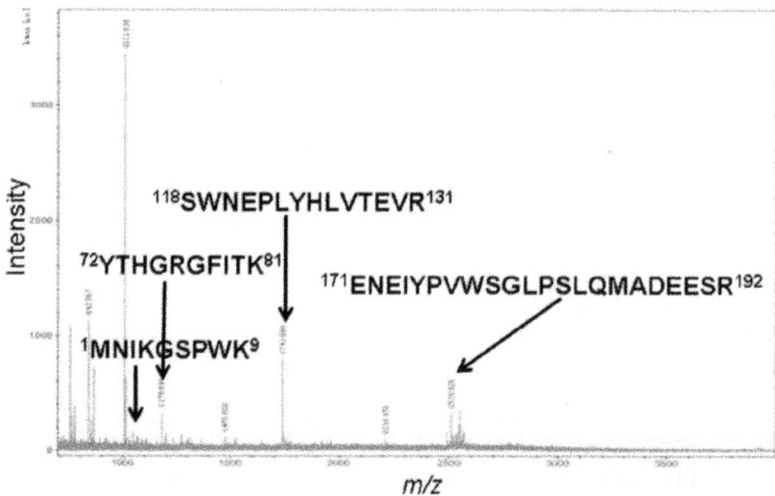

Figure 3.
PMF analysis of hPRL that was contained in spot v6. Reproduced from Qian et al. [9], with copyright permission from Frontiers in open access article, copyright 2018.

with NetPhos 3.1 Server with a score more than 0.5. It obtained 22 statistically significantly phosphorylation sites in hPRL (position 1–227). N-glycosylation sites in hPRL (position 1–227) were predicted with NetNGlyc 1.0 Server with score more than 0.5. It obtained 10 statistically significant N-glycosylation sites in hPRL (position 1–227). O-glycosylation sites in hPRL (position 1–227) were predicted with NetOGlyc 4.0 Server with score more than 0.5. It obtained six statistically significant O-glycosylation sites in hPRL (position 1–227). These data suggest that PTMs such as phosphorylation and glycosylation might be the important reason to cause the PRL proteoforms.

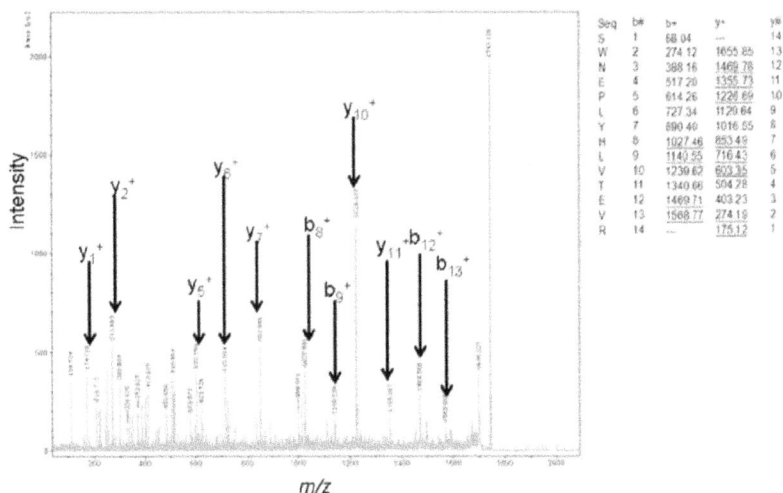

Seq	b#	b+	y+	y#
S	1	68.04	—	14
W	2	274.12	1655.85	13
N	3	388.16	1469.78	12
E	4	517.20	1355.73	11
P	5	614.26	1226.69	10
L	6	727.34	1129.64	9
Y	7	890.40	1016.55	8
H	8	1027.46	853.49	7
L	9	1140.55	716.43	6
V	10	1239.62	603.35	5
T	11	1340.66	504.28	4
E	12	1469.71	403.23	3
V	13	1568.77	274.19	2
R	14	—	175.12	1

Figure 4.
MS/MS analysis of the tryptic peptide 118SWNEPLYHLVTEVR131 that was derived from PRL in spot v6. Reproduced from Qian et al. [9], with copyright permission from Frontiers in open access article, copyright 2018.

Calc. [M + H]⁺	Position	Characteristic tryptic peptide sequence	Observed [M + H]⁺
505.2803	1–4	MNIK	–
1060.5608	1–9	MNIKGSPWK	+
3930.1893	1–38	MNIKGSPWKGSLLLLLVSNLLLCQSVAPLPICPGGAAR	–
574.2983	5–9	GSPWK	–
3443.9268	5–38	GSPWKGSLLLLLVSNLLLCQ SVAPLPICPGGAAR	–
2888.6463	10–38	GSLLLLLVSNLLLCQSVAPL PICPGGAAR	–
3589.0154	10–44	GSLLLLLVSNLLLCQSVAPL PICPGGAARCQVTLR	–
954.5189	29–38	LPICPGGAAR	–
1654.8879	29–44	LPICPGGAARCQVTLR	–
2301.1954	29–49	LPICPGGAARCQVTLRDLFD R	–

+, this peptide ion was observed with mass spectrometry in each MS spectrum.
–, this peptide was not observed with mass spectrometry.
Reproduced from Qian et al. [9], with copyright permission from Frontiers in open access article, copyright 2018.

Table 2.
Characteristic tryptic peptides to determine signal peptide (position1–28) within human PRL proteoforms in human pituitary.

3.5 Potential clinical application of PRL proteoform pattern

Prolactin synthesized in the ribosome in the pituitary secretes into blood circulation to reach its target organ and exert its biological roles, which is closely associated with multiple physiological and pathological processes, including pituitary adenomas. This study found six PRL proteoforms with different with differential isoelectric point (p*I*) and relative mass (*M*ᵣ) in control pituitary tissues, which were identified with 2DGE coupled with Western blot and MS. This prolactin

PRL proteoform no.	Swiss-Prot no.	pI	M_r	Ratio (NF⁻: Con)	Ratio (FSH⁺/ LH⁺: Con)	Ratio (FSH⁺: Con)	Ratio (LH⁺: Con)	Ratio (PRL: Con)
V1	P01236	6.1	26.0	−8.3	−99.9	−46.2	−12.6	−3.4
V2[a]	P01236	6.3	26.4	−4.9	−3.8	1	−4.1	1
V3	P01236	6.3	27.9	1	−12.3	−14.6	−26.2	1
V4[b]	P01236	6.5	26.1	−100	−19.0	−17.6	−20.1	1
V5	P01236	6.8	25.9	−100	−19.7	−100	−36.7	1
V6	P01236	6.7	25.9	−100	−32.6	−11.3	−33.6	1

Modified from Qian et al. [9], with copyright permission from Frontiers in open access article, copyright 2018.
[a]*Characterized with LC-ESI MS/MS.*
[b]*Characterized with LC-ESI-MS/MS and MALDI-TOF PMF.*
All other proteins were characterized with MALDI-TOF PMF. Con, control; −, decreased relative to controls; −100, lost relative to controls; 1, no change relative to controls; M_r, kDa.

Table 3.
Prolactin proteoform pattern changed in different subtypes of pituitary adenomas compared to control pituitaries.

PRL proteoform	NF⁻ (%)	FSH⁺/LH⁺ (%)	FSH⁺ (%)	LH⁺ (%)	PRL⁺ (%)
V1	2.64	53.34	24.24	9.45	40.48
V2	1.56	2.03	0.52	3.08	11.90
V3	0.31	6.57	7.66	19.65	11.91
V4t	31.83	10.14	9.18	15.08	11.90
V5	31.83	10.52	52.47	27.53	11.91
V6	31.83	17.40	5.93	25.21	11.90
Total	100.00	100.00	100.00	100.00	100.00

Reproduced from Qian et al. [9], with copyright permission from Frontiers in open access article, copyright 2018.
Chi-square test = 360.606, p = 0.000 (p < 0.01) among five subtypes of pituitary adenomas.

Table 4.
Proportional ratio changes of PRL proteoforms among five subtypes of pituitary adenomas.

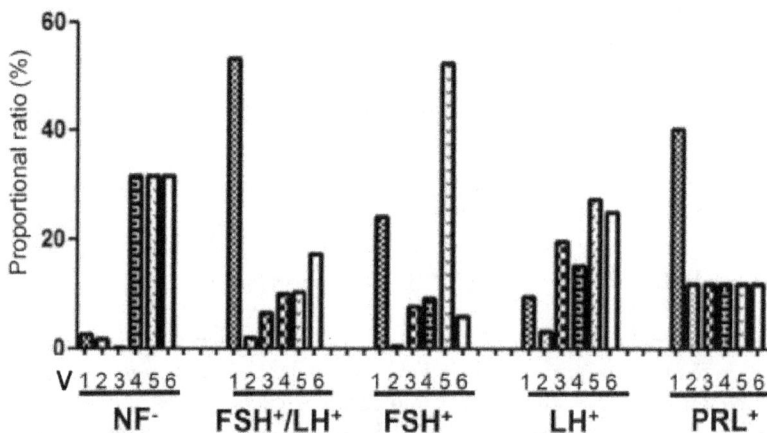

Figure 5.
Proportional ratio changes of PRL proteoforms among five subtypes of pituitary adenomas. Reproduced from Qian et al. [9], with copyright permission from Frontiers in open access article, copyright 2018.

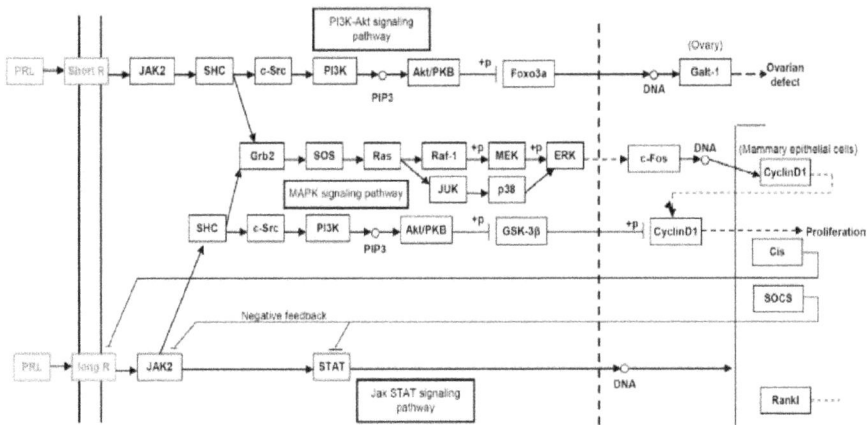

Figure 6.
PRL proteoform-driven signaling pathway via the short or long PRL receptors. Reproduced from Qian et al.
[9], with copyright permission from Frontiers in open access article, copyright 2018.

proteoform pattern was significantly changed among different hormone-subtypes of nonfunctional pituitary adenoma (NF⁻, LH⁺, FSH⁺, and LH⁺/FSH⁺) and prolactinoma (PRL⁺) tissues. This result suggests the potentially important clinical value of serum PRL proteoforms. The reason is that pituitary tissues are impossible to obtain for clinical diagnosis, and prolactin must secrete into blood to exert its biological roles, we strongly believe serum PRL proteoforms exist and the serum PRL proteoform pattern changes among different pituitary adenomas. Therefore, we will further analyze serum PRL proteoform pattern changes among different subtypes of pituitary adenomas, and develop the PRL proteoform pattern as biomarker for prediction, diagnosis, or prognostic assessment of pituitary adenoma occurrence, progression, and prognosis.

4. Conclusions

Six PRL proteoforms were identified in human pituitary tissue with 2DGE and MS analyses, and four of six PRL proteoforms were validated with 2DGE-based Western blot, MS, and MS/MS analyses. There were significant differences in PRL proteoform pattern among five different subtypes of pituitary adenomas (LH⁺, NF⁻, FSH⁺, FSH⁺/LH⁺, and PRL⁺) ($P < 0.05$). Moreover, MS analysis revealed that six PRL proteoforms are PRL prohormone. PRL proteoforms might be derived from PTMs such as phosphorylation, deamidation, and glycosylation. Further, different PRL proteoforms might bind to different PRL receptors to produce different physiological functions. These findings provide scientific basis for in-depth understanding of pituitary adenomas, and help develop biomarkers for treatment of pituitary adenoma patients. The serum PRL proteoform pattern has important clinical application value for prediction, diagnosis, and prognostic assessment of pituitary adenomas.

Acknowledgements

The authors acknowledge the financial supports from the Hunan Provincial Hundred Talent Plan (to X.Z.), National Natural Science Foundation of China

(Grant No. 81572278 and 81272798 to X.Z.), China "863" Plan Project (Grant No. 2014AA020610-1 to X.Z.), the Hunan Provincial Natural Science Foundation of China (Grant No. 14JJ7008 to X.Z.), and the Xiangya Hospital Funds for Talent Introduction (to X.Z.).

Conflict of interest

The authors declare that they have no financial and personal relationships with other people or organizations.

Author's contributions

X.Z. conceived the concept, designed the manuscript, wrote and critically revised the manuscript, coordinated, and was responsible for the correspondence work and financial support. Q.S. participated in the literature analysis, data analysis, prepared figures, and wrote partial manuscript.

Abbreviations

DTT	dithiothreitol
FSH	follicle-stimulating hormone
GO	gene ontology
IEF	isoelectric focusing
IPG	immobilized pH gradient
LH	luteinizing hormone
M_r	molecular weight
MS	mass spectrometry
pI	isoelectric point
PRL	prolactin
PTM	post-translational modifications
PVDF	polyvinylidene fluoride
SDS-PAGE	sodium dodecyl sulfate polyacrylamide gel electrophoresis
2DGE	two-dimensional gel electrophoresis

Author details

Xianquan Zhan[1,2,3*] and Shehua Qian[1,2,3]

1 University Creative Research Initiatives Center, Shandong First Medical University, Jinan, Shandong, China

2 Key Laboratory of Cancer Proteomics of Chinese Ministry of Health, Xiangya Hospital, Central South University, Changsha, China

3 State Local Joint Engineering Laboratory for Anticancer Drugs, Xiangya Hospital, Central South University, Changsha, China

*Address all correspondence to: yjzhan2011@gmail.com

IntechOpen

References

[1] Kobayashi T, Usui H, Tanaka H, Shozu M. Variant prolactin receptor in agalactia and hyperprolactinemia. The New England Journal of Medicine. 2018;**379**(23):2230-2236. DOI: 10.1056/NEJMoa1805171

[2] Owerbach D, Rutter WJ, Cooke NE, Martial JA, Shows TB. The prolactin gene is located on chromosome 6 in humans. Science (New York, N.Y.). 1981;**212**(4496):815-816

[3] Larsen CM, Grattan DR. Prolactin, neurogenesis, and maternal behaviors. Brain, Behavior, and Immunity. 2012;**26**(2):201-209. DOI: 10.1016/j.bbi.2011.07.233

[4] de la Fuente JR, Rosenbaum AH. Prolactin in psychiatry. The American Journal of Psychiatry. 1981;**138**(9):1154-1160. DOI: 10.1176/ajp.138.9.1154

[5] Binart N, Bachelot A, Bouilly J. Impact of prolactin receptor isoforms on reproduction. Trends in Endocrinology and Metabolism: TEM. 2010;**21**(6):362-368. DOI: 10.1016/j.tem.2010.01.008

[6] Xie W, Liu H, Liu Q, Gao Q, Gao F, Han Y, et al. Seasonal expressions of prolactin, prolactin receptor and STAT5 in the scented glands of the male muskrats (*Ondatra zibethicus*). European Journal of Histochemistry: EJH. 2019;**63**(1). DOI: 10.4081/ejh.2019.2991

[7] Bernichtein S, Touraine P, Goffin V. New concepts in prolactin biology. The Journal of Endocrinology. 2010;**206**(1):1-11. DOI: 10.1677/joe-10-0069

[8] Ben-Jonathan N, LaPensee CR, LaPensee EW. What can we learn from rodents about prolactin in humans?

Endocrine Reviews. 2008;**29**(1):1-41. DOI: 10.1210/er.2007-0017

[9] Qian S, Yang Y, Li N, Cheng T, Wang X, Liu J, et al. Prolactin variants in human pituitaries and pituitary adenomas identified with two-dimensional gel electrophoresis and mass spectrometry. Frontiers in Endocrinology. 2018;**9**:468. DOI: 10.3389/fendo.2018.00468

[10] Zhan X, Giorgianni F, Desiderio DM. Proteomics analysis of growth hormone isoforms in the human pituitary. Proteomics. 2005;**5**(5):1228-1241. DOI: 10.1002/pmic.200400987

[11] Freeman ME, Kanyicska B, Lerant A, Nagy G. Prolactin: Structure, function, and regulation of secretion. Physiological Reviews. 2000;**80**(4):1523-1631. DOI: 10.1152/physrev.2000.80.4.1523

[12] Schaffer LV, Millikin RJ, Miller RM, Anderson LC, Fellers RT, Ge Y, et al. Identification and quantification of proteoforms by mass spectrometry. Proteomics. 2019;**19**(10):e1800361. DOI: 10.1002/pmic.201800361

[13] Oliveira BM, Coorssen JR, Martins-de-Souza D. 2DE: The phoenix of proteomics. Journal of Proteomics. 2014;**104**:140-150. DOI: 10.1016/j.jprot.2014.03.035

[14] Zhan X, Yang H, Peng F, Li J, Mu Y, Long Y, et al. How many proteins can be identified in a 2DE gel spot within an analysis of a complex human cancer tissue proteome? Electrophoresis. 2018;**39**(7):965-980. DOI: 10.1002/elps.201700330

[15] Kurgan N, Noaman N, Pergande MR, Cologna SM, Coorssen JR, Klentrou P. Changes to the human serum proteome in response

to high intensity interval exercise: A sequential top-down proteomic analysis. Frontiers in Physiology. 2019;**10**:362. DOI: 10.3389/fphys.2019.00362

[16] Zhan X, Huang Y, Long Y. Two-dimensional gel electrophoresis coupled with mass spectrometry methods for an analysis of human pituitary adenoma tissue proteome. Journal of Visualized Experiments: JoVE. 2018;**134**:e56739. DOI: 10.3791/56739

[17] Li C, Zhan X, Li M, Wu X, Li F, Li J, et al. Proteomic comparison of two-dimensional gel electrophoresis profiles from human lung squamous carcinoma and normal bronchial epithelial tissues. Genomics, Proteomics & Bioinformatics. 2003;**1**(1):58-67

[18] Wang K, Lin Z, Zhang H, Zhang X, Zheng X, Zhao L, et al. Investigating proteome and transcriptome response of *Cryptococcus podzolicus* Y3 to citrinin and the mechanisms involved in its degradation. Food Chemistry. 2019;**283**:345-352. DOI: 10.1016/j.foodchem.2019.01.052

[19] Kadi AA, Yin W, Rahman A. In-vitro metabolic profiling study of potential topoisomerase inhibitors 'pyrazolines' in RLMs by mass spectrometry. Journal of Chromatography B, Analytical Technologies in the Biomedical and Life Sciences. 2019;**1114-1115**:125-133. DOI: 10.1016/j.jchromb.2019.03.026

[20] Zhan X, Li N, Zhan X, Qian S. Revival of 2DE-LC/MS in proteomics and its potential for large-scale study of human proteoforms. Med One. 2018;**3**(5):e180008. DOI: 10.20900/mo.20180008

[21] Moreno CS, Evans CO, Zhan X, Okor M, Desiderio DM, Oyesiku NM. Novel molecular signaling and classification of human clinically nonfunctional pituitary adenomas identified by gene expression profiling and proteomic analyses. Cancer Research. 2005;**65**:10214-10222. DOI: 10.1158/0008-5472.can-05-0884

[22] Evans CO, Moreno CS, Zhan X, Mccabe MT, Vertino PM, Desiderio DM, et al. Molecular pathogenesis of human prolactinomas identified by gene expression profiling, RT-qPCR, and proteomic analyses. Pituitary. 2008;**11**:231-245. DOI: 10.1007/s11102-007-0082-2

[23] Zhan X, Desiderio DM. Mass spectrometric identification of in vivo nitrotyrosine sites in the human pituitary tumor proteome. Methods in Molecular Biology. 2009;**566**:137-163. DOI: 10.1007/978-1-59745-562-6_10

[24] Zhan X, Desiderio DM. A reference map of a human pituitary adenoma proteome. Proteomics. 2003;**3**:699-713. DOI: 10.1002/pmic.200300408

[25] Peng F, Li J, Guo T, Yang H, Li M, Sang S, et al. Nitroproteins in human astrocytomas discovered by gel electrophoresis and tandem mass spectrometry. Journal of the American Society for Mass Spectrometry. 2015;**26**:2062-2076. DOI: 10.1007/s13361-015-1270-3

[26] Blom N, Sicheritz-Ponten T, Gupta R, Gammeltoft S, Brunak S. Prediction of post-translational glycosylation and phosphorylation of proteins from the amino acid sequence. Proteomics. 2004;**4**:1633-1649. DOI: 10.1002/pmic.200300771

[27] Blom N, Gammeltoft S, Brunak S. Sequence and structure-based prediction of eukaryotic protein phosphorylation sites. Journal of Molecular Biology. 1999;**294**:1351-1362. DOI: 10.1006/jmbi.1999.3310

[28] Steentoft C, Vakhrushev SY, Joshi HJ, Kong Y, Vester-Christensen MB,

Schjoldager KT, et al. Precision mapping of the human O-GalNAc glycoproteome through SimpleCell technology. The EMBO Journal. 2013;**32**:1478-1488. DOI: 10.1038/emboj.2013.79

[29] Gupta R, Brunak S. Prediction of glycosylation across the human proteome and the correlation to protein function. Pacific Symposium on Biocomputing. 2002;**2002**:310-322. DOI: 10.1142/9789812799623_0029

Proteoforms in Acute Leukemia: Evaluation of Age- and Disease-Specific Proteoform Patterns

Fieke W. Hoff, Anneke D. van Dijk and Steven M. Kornblau

Abstract

Acute leukemia are a heterogeneous group of malignant diseases of the bone marrow that occur at all ages. Acute lymphoid leukemia (ALL) accounts for about 80% of all pediatric leukemia patients, whereas acute myeloid leukemia (AML) is more common in adults compared to pediatric patients. Despite similar patterns in the pathogenesis of acute leukemia in children and adults, clinical outcome in response to therapy differs substantially. Studying proteoforms in acute leukemia in children and adults, might identify similarities and differences in crucial signaling pathways that play a key role in the development or progression of the disease. In this chapter we will discuss how the study of proteoforms in acute leukemia could potentially contribute to a better understanding of the leukemogenesis, can help to identify effective targets for specific targeted treatment approaches in different subgroups of age and disease, and could aid the development of reliable biomarkers for prognostic stratification.

Keywords: acute myeloid leukemia, acute lymphoblastic leukemia, proteoforms, RPPA, pediatrics

1. Introduction

Acute leukemia forms a group of rapidly progressing malignant diseases characterized by a block in the differentiation and an uncontrolled clonal proliferation of abnormal hematopoietic progenitor cells in the bone marrow and the peripheral blood [1, 2]. This accumulation of immature cells ("blasts") interferes with the production of normal blood cells, causing neutropenia, thrombocytopenia and anemia. According to the lineage of origin of the progenitor cells, the common lymphoid or the common myeloid, acute leukemia can roughly be classified into acute lymphoblastic leukemia (ALL) and acute myeloid leukemia (AML).

Acute leukemia patients are diagnosed using morphologic, cytochemical and immunophenotypic methods, and are further sub-classified by chromosomal analysis and the presence or absence of somatically acquired gene mutations. While classification allows for prediction of outcome, the outcome risk of a large group of patients is still difficult to define. In addition, treatment options are expanding that treat patients based on their genetic abnormalities (in particular in the adult population), but so far most genetic abnormalities are not yet targetable, and most drugs that enter clinical trials rely on the increased abundance or altered activity of proteins, namely specific proteoforms, instead of the genetic lesion itself.

Proteoforms are defined as different forms of a protein derived from a single gene, and include all forms of genetic variation (e.g. amino acid variation), alternative splicing, and post-translational modifications (PTM). This means that one transcribed gene can lead to a variety of protein structures, and that the biological function of the final proteoform, as well as the cellular localization, binding partners and kinetics can vary greatly. As this suggests that gene sequences do not accurately predict the expression of a protein or whether the protein is stable or functional, it is not surprising that transcriptome data only correlates for about 17–40% with protein abundance [3–5]. Proteoforms are the basic units of a proteome. We believe that the study of proteoforms is an essential strategy to reveal cell dependencies and their underlying mechanism, and that this could add in the process of risk stratification and could identify novel therapeutic targets in highly complex diseases such as acute leukemia. Moreover, as the cure rates between ALL and AML, and between children and adults markedly differ, a direct comparison of the leukemic proteoforms between those patients, may aid to unravel the biological pathogenesis, and reveal similarities and dissimilarities that can propose therapeutics that target these proteoforms in one disease, that could also be effective in an otherwise disparate leukemia that shares protein patterns.

2. Acute leukemia

2.1 Acute lymphoblastic leukemia

ALLs are neoplasms composed of immature B (pre-B), T (pre-T) or NK-cells that are referred as ("lymphoblasts"), of which the majority is pre-B ALL (85–90% in children vs. 75% in adults) [1]. It is the most common cancer in children and accounts for a quarter of all childhood malignancies. Although there are as many adults with ALL as there are children with the disease, the relative frequency in adults is much lower. Worldwide, the overall incidence is approximately 1–2 per 100.000 people, with a peak incidence occurring in childhood and a second peak above the age of 50 years [6].

2.1.1 Cytogenetic abnormalities

Chromosomal aberrations are the hallmark of ALL and are often used to categorize patients. In B-ALL, recurrent chromosomal abnormalities are found in 80% of the patients, including numerical and structural changes as translocations, deletions and inversions. There are substantial differences in the frequencies of occurring of cytogenetic abnormalities between children and adults [1, 7–10]. For instance, the translocation 9;22 [BCR-ABL1] is observed in 2–5% of the children compared to in 30% of the adults, whereas the translation 12;21 [ETV6-RUNX1] is observed in 25% of the pediatric patients versus 3% in adult population. The hyperdiploid (gain of chromosomes) karyotype is present in 30–40% of the children compared to 3% in adults. Finally, translocation 4;11 resulting in the MLL-AF4 fusion gene, is detected in 60–80% of the infants (younger than 1 year old), whereas it is seen in only 2% of the patients up to 15 years and rare in adults. Hypodiploidy (loss of chromosomes) occurs in 5–6% of the ALL patients, independent of age.

Chromosomal translations occur less frequently in T-ALL compared to B-ALL (approximately 50–60%) and unlike in B-ALL their prognostic impact is not well defined and they are not used for risk stratification [10]. They are involved in both the T-cell receptor and the non-T-cell receptor loci on the chromosome or aberrant expression of the transcription factor oncogenes. There is less association with age [8].

2.1.2 Prognosis

Survival in children with ALL is much better compared to adults, with the exception of infant ALL. In 60–80% of the cases, infant ALL is characterized by translocations involving 11q23, affecting the KMT2A gene. Aberrant KMT2A in ALL is associated with a high rate of early treatment failure and a very poor outcome (long-term event-free survival of 28–45%), even when treated with more aggressive chemotherapy regimens [11, 12]. Historically, pediatric T-ALL was considered as high-risk disease. With the introduction of therapy intensification in T-cell ALL, this has changed to outcomes comparable to B-cell ALL, resulting in a five-year OS rate of more than 90% [13]. However, within certain high-risk subgroups (e.g. infants or children ≥10 years of age), 25–30% still experience relapse, which has a dismal outcome even with hematopoietic stem cell transplantation. Death resulting from treatment toxicity remains a challenge with an estimated 10-year cumulative incidence of treatment-related death of 2.9% [14].

Survival for ALL in adults is around 45%, but patients above the age of 60 suffer from inferior outcomes with only 10–15% long-term survival [10]. This is, at least partially, due to higher risk of medical comorbidities, the inability to tolerate standard chemotherapy regimens, and age-related unfavorable intrinsic biology such as Philadelphia chromosome positive, hypodiploidy and complex karyotype. However, as even the adolescents and young adults who lack medical comorbidities do significantly worse compared to their younger counterparts, the contribution of the different underlying biology should not be underestimated [8].

2.2 Acute myeloid leukemia

In general, patients with AML have similar signs and symptoms as patients with ALL which mainly includes symptoms related to (pan)cytopenia. AML is the most common acute leukemia in adults, whereas it is relatively rare in children (accounting for only 10% of the acute leukemia) [15]. Overall, AML occurs in 3–5 cases per 100.000 people, and the incidence strongly increases with age.

2.2.1 Cytogenetic and molecular abnormalities

AML is a very heterogeneous disease and the identification of AML-associated chromosomal translocations and inversions have led to the current 2016 World Health Organization (WHO) classification system [16]. In this classification, eight recurrent genetic abnormalities (e.g. translocation (15;17) [*PML-RARA*], translocation (8;21) [*RUNX1-RUNX1T1*], inversion (16) or translocation (16;16) [*CBFB-MYH1*], translocation (9;11) [*MLLT3-KMT2A*], and translocation (9;22) [*BCR-ABL1*]) and their variants are included. In approximately 50 percent of patients, no cytogenetic abnormalities will be present, referred to as "normal karyotype" [17]. Additional classification in AML is provided by detection of one or more recurrent genetic mutations, with *NPM1*, *FLT3*, *IDH1*, *IDH2*, *RUNX1* and *CEBPA* most studied.

Recently, the Therapeutically Applicable Research to Generate Effective Treatments (TARGET) study has presented the molecular landscape of nearly 1000 pediatric AML patients that participated in several Children's Oncology Group (COG) clinical trials [18]. Like adult AML, they found that pediatric AML has one of the lowest rates of mutations as compared to other cancers as recognized by The

Cancer Genome Atlas, suggesting that the number of recognized recurrent mutations in AML alone is not sufficient to explain its heterogeneity. They demonstrated
that the landscape of somatic variants in pediatric AML was markedly different
from that reported in adults, highlighting the need for and facilitate the development of age-tailored targeted therapies for the treatment of pediatric AML [19, 20].

2.2.2 Prognosis

Among adult patients who are under 60 years of age, AML can be cured in
35–40% of the patients, whereas the survival rates of patients older than 60 is only
5–15%. For older patients who are unable to receive intensive chemotherapy without
acceptable side effects the prognosis is even more dismal, with a median survival
of only 5–10 months [2]. Survival rates in the pediatric population, have improved
greatly, although OS rates of 65–70% are still much lower than that for pediatric
ALL [21].

3. Proteome differs from transcriptome

The human genome is the total amount of DNA that each cell in the body
contains, including an estimated of 30,000–40,000 protein-coding genes. While
the basic dogma of biology formerly was that DNA was transcribed into messenger
RNA, which is then translated into proteins, and that mRNA levels could be used to
predict protein abundance, it becomes more and more clear that this is overly sim-
plistic due to our expanding knowledge of the effects of epigenetics, environmental
influences, mRNA editing, alternative splicing and noncoding RNAs on gene
expression. For instance, coding single-nucleotide polymorphisms and mutations
can affect the final protein sequence and function, and based on endogenous prote-
olysis and mRNA splicing, different isoforms can be generated from the same set of
nucleotides. Additionally, after translation of the RNA transcript, proteins undergo
multiple modifications affecting the protein function, localization, lifespan and
activity. Together this results in up to a million of proteoforms.

One of the first studies back in 1999, that compared a limited number of mRNA
and proteins using *Saccharomyces cerevisiae*, already concluded that the correlation
between both was only 0.36 [4, 22]. And, even with the significant improvements in
high-throughput genomic and proteome approaches, this fundamental observation
continues to be widely, though not universally, supported, as most studies nowadays
still show a correlation coefficient that varies between 0.17 and 0.40. Per example,
Mun et al. recently performed correlation analysis of mRNA and protein log2-fold
changes between gastric cancer tumor samples and adjacent normal tissues using
6803 genes with protein and mRNA abundances available in at least 30% ($\geq$24) of
the patients. Of the 6803 genes, only 34.3% showed significant (FDR < 0.01) positive
correlation with an average correlation coefficient of 0.28 [23]. Zang et al., performed an integrated proteogenomic analyses human colon and rectal cancer samples
and while 89% of the samples showed significant positive mRNA-protein correlation
(of which only 32% was significantly correlated), the average correlation between
messenger RNA transcript abundance and protein abundance was only 0.23 [24].

4. Age-associated proteoforms in acute leukemia

Aforementioned, the functional variant of a protein, the proteoform, is defined
by genetics, mRNA editing, and PTMs. In particular in ALL, that peaks between 2

and 5 years of age followed by a gradual increase in the older patients, is it suggested that different combinations of genetic factors (resulting in different proteoforms) contribute to leukemogenesis at different ages. In order to answer the question why genes are differentially expressed upon age, a closer look at biological processes that influence the final proteoform production via pre-translational modifications may help. Here we will discuss a few examples of how these differ between younger and older patients with acute leukemia.

4.1 Genetic variants

Emerging genome wide sequencing techniques identified disease and age-specific gene variants in acute leukemia. For example, Perez-Andreu et al. discovered a single nucleotide polymorphism (SNP), a variant of the coding region of the DNA, of *GATA3* on 10p14 that was associated with the susceptibility to ALL in adolescents and young adults, and that progressively increased with age [25]. Furthermore, genomic variants that occur in both pediatric and adult leukemia sometimes display a different phenotype at the protein level. As shown by Zuurbier et al., loss of *PTEN* protein due to the production of an unstable and truncated proteoform caused by a frameshift mutation or genomic deletion is a frequently seen in T-cell ALL (predominantly in pediatric T-ALL). *PTEN* is often recognized as a tumor suppressor, but its behavior and relation to outcome is highly context dependent. *PTEN* abnormalities may impact *NOTCH1* and, in a cohort of *PTEN* mutated pediatric T-ALL patients (with loss of *PTEN* protein) that lacked the *NOTCH1* activating mutations, had significantly fewer relapses compared to patients with activated *PTEN* and *NOTCH1* [26]. In contrast, another study showed that *PTEN* mutations without *NOTCH1* abnormalities were associated with poor prognosis in adults [37]. Thus, genomic mutations within the same gene, do not always produce the same proteoform with the same function. Mutations can create a proteoform with a completely different function and can convert a protein from a tumor suppressor into a tumor driver [27]. Although, genome wide studies are very meaningful in detection of conditions specific to age and disease, but the net effect on the cell largely depends on the production of the final proteoform (tumor suppressor or tumor driver) and the pathways they act in.

4.2 Chromosomal translocations

Chromosomal abnormalities, gene fusions and copy number aberrations are more common in the younger patient population [28]. The ratio of structural variation to mutational burden decreases continuously with age, with the most chromosomal translocations in infants (<1 year) compared to all other ages. Within this young age-group, the most common fusion involves *KMT2A* (also known as *MLL1*), present in 38% of the infants [28]. A second age-peak is recognized in young to middle aged AML adults. Overall, more than 80 fusion partners of KMT2A are described and it is the protein partner of KMT2A that determine characteristics specific to age and disease. Interestingly, 50% of the infants younger than 1 year with ALL contain the specific MLL-AF4 fusion protein caused by the t(4;11)(q21,q23) translocation [29], whereas in AML, the most common MLL rearrangement is the MLL-AF9 that arises from a t(9;11)(p22,q23) translocation. In both populations, MLL leukemia confers poor prognosis and identification of unique proteoforms in this subtype leukemia may guide treatment stratification by providing targetable leads. For instance, downstream proteomic targets mediated by MLL-AF4 include *HOX*, *EPHA7*, *MEIS*, *PBX* and *GSK-3* and these are already considered or investigated as therapeutic targets in the context of MLL-rearranged leukemia. In addition, *RAS*, *DOT1L*, and *HSP-90* also have been described as potential targets in MLL leukemia [30]. As those genes and their protein products are in particular involved in transcription regulation, we hypothesize that patients that expression differences in those conserved genes, likely also harbor differences in abundance, or proteoforms of its downstream proteins, compared to wild-type patients.

4.3 Non-coding microRNAs

MicroRNAs (miRNA) are small non-coding RNAs that affect the proteome through their binding to mRNA influencing/inhibiting the translation to proteins. Aberrant miRNA expression is associated with leukemogenesis [31], and multiple miRNAs are found to be expressed differently upon age. A study by Noren Hooten et al. showed downregulation of miRNA expression in peripheral blood of healthy individuals with advancing age. Cancer is often age-related and five out of nine downregulated miRNAs in this study were related to cancer pathogenesis [32]. Another study compared miRNA profiles between pediatric and adult patients with AML and again, identified significant lower miRNA expression in adults compared to children. In addition, they found distinct miRNA expression patterns in both t(8;21) and t(15;17) translocated pediatric AML, but not in adults. Also, nine-fold upregulation of miR-21 was identified in the MLL-rearranged pediatric patients compared to others and this finding was also not reflected by the MLL-rearranged adult population [33]. The identification of age-specific miRNA specific expressing in leukemia together with the fact that miRNA will affect the final proteomic state, indicates that further proteomic approaches could likely unravel differences in proteoforms between younger and older patients within leukemic subtypes.

4.4 Post-translational modifications

DNA is wrapped around histone to form a compact chromatin structure and PTMs on histone tails, such as the addition or removal of methyl or acetyl groups on lysine residues, or direct DNA methylation regulate chromatin accessibility and initiate and maintain gene expression patterns that account for specific cell lineage differentiation and development [34]. Packaging of the chromatin structure changes with age and include global loss of heterochromatin resulting in a more open chromatin state in the elderly. Reduction of heterochromatin due to increased histone acetylation during aging is also well-established [35, 36], but less well-characterized is the role of histone methylation. Since the prevalence of AML increases with age, we asked ourselves if histone methylation profiles are different between pediatric and adult AML. We recently applied RPPA-based profiling using antibodies against multiple histone methylation sites which enabled us to define disease and age characteristic patterns of histone modification. In agreement with our hypothesis, a significant decline in histone methylation was seen upon age in both ALL and AML cases (manuscript in preparation).

As mentioned, MLL-rearrangements are specific to age and disease, and are frequently altered in leukemia. As MLL fusion proteins modulate the chromatin structure by histone tail modifications, MLL-rearranged leukemia is considered as epigenetic malignancy. In addition, mutations in proteins that modify the histone PTM process (e.g. writers, erasers and readers) are more frequently found in T-ALL compared to other childhood malignancies, and distinct DNA methylation patterns were recognized among different subtypes of ALL. Those patterns correlated with changed transcriptomes. Aberrant DNA methylation is associated with silencing of genes that involved in lymphoid development, and contribute to leukemogenesis. By combining DNA methylation and transcriptome analysis, transcriptional silencing via promotor hypermethylation was recently identified in pediatric AML [28], and correlated with age, karyotype and outcome.

Since hypomethylating agents have been widely used to treat, in particular the older leukemia patients, we hypothesize that proteomics can help to identify more refined subgroups (maybe even from the younger population) that can be treated with certain treatment regimens that alter the epigenome. For instance, the discovery of a specific protein or proteomic signature (either related to the epigenome or not) that is correlated with sensitivity to hypomethylating agents, can potentially act as biomarker in MLL-rearranged leukemia, to select patients that can benefit from those agents. In the experimental setting, therapies with hypomethylating agents have already showed re-expression of the hypermethylated genes along with restored chemosensitivity, and in relapsed ALL, increased promotor methylation was found to be related to increased chemoresistance [37]. If it is possible to identify a set of proteins that is specific for relapsed MLL-rearrangement AML and/or ALL, rapid tests (e.g. ELISA, IHC or FPPA) could be developed to quickly provide information about the protein abundance in relapsed patients, to identify those who will benefit from additional treatment, as well as, *a priori*, predict which newly diagnosed patients are most likely to relapse, and treat those with additional treatment to prevent relapse.

5. Oncogenic proteoforms leading to leukemia

Mutations in the DNA of the hematopoietic stem cells play a pivotal role in leukemogenesis and within single genes, multiple mutations have been identified that results in different forms of the protein. One example involves transcription factor CCAAT/enhancer binding protein A (CEBPA) mutated AML patients, which is known to regulate growth arrest and differentiation in hematopoiesis by promoting granulocyte lineage differentiation in common myeloid progenitor cells, and disruption of normal CEBPA expression in myeloid progenitors may lead to a block in granulopoiesis resulting in erythropoiesis in its place [38]. As critical regulator of myeloid lineage development it is not surprising that CEBPA is mutated in ~10% of AML patients and most frequently classified as myeloblastic AML subtype M1 of M2 according the French-American-British (FAB) classification. CEBPA transcript translates for a full-length (CEBPA-p42) or shorter isoform (CEBPA-p30). CEBPA-p30 isoforms contain the DNA binding domain but lack the N-terminal transactivation domain. However, CEBPA-p30 is dominant negative by reducing transcriptional activity after heterodimerization with full-length CEBPA-p42. About half of CEBPA mutated AML patients have one allele with a N-terminal mutation and one allele with a C-terminal mutation. The N-terminal mutant results in translational termination of the full-length isoform and increase truncated CEBPA-p30 expression. In contrast, C-terminal mutations in CEBPA-p42 are mostly characterized by in-frame basic region leucine zipper (bZIP) variants inhibiting normal CEBPA function by disrupting DNA binding and dimerization [39]. CEBPA mutated patients might be candidates for inhibition of the oncogenic CEBPA-p30 isoform to recover the disrupted p42/p30-ratio.

6. High-throughput proteomics methodologies

Proteomics may be the least developed and investigated "-omics" approach, it is likely one of the most informative for understanding of cellular behavior as it can provide useful information about both protein abundance and activity, as regulated by the PTM, the protein-protein and protein-DNA interactions. Nowadays, two of the most commonly used high-throughput techniques to study the proteome in leukemia are mass-spectrometry (MS)-based techniques and antibody-based techniques.

6.1 MS-based

MS is a high-throughput technique uses the formation of ions (charged fragments) from the protein analyte to distinguish between proteoforms. Those ions can be sorted and measured using electrical and/or magnetic fields based on their mass-to-charge ratio (m/z), and identification of the protein follows based on the abundance of those m/z-fragments [40]. Globally, proteins can be ionized with two distinct methods: matrix assisted laser desorption/ionization (MALDI) and electrospray ionization (ESI). In MALDI the protein sample is mixed with an energy absorbing matrix. Irradiation of this matrix causes vaporization of the matrix together with the sample, resulting in the formation of ions [41]. ESI creates ions using electrospray to dissolve the protein lysate, by applying high-voltage to the dissolvent to create an aerosol of small charged fragments. When a protein sample is highly complex, samples may require separation prior to MS analysis using 1D or 2D gel electrophoresis, high-pressure liquid chromatography (LC-MS), or gas chromatography (GC-MS) to maximize the sensitivity. Because proteoforms are derived from a single gene, they often contain homologous sequence regions, and because of the digestion step, information about the relationship between amino acid sequence and the PTM often lacks, this significantly complicates the process of identifying proteoforms. Several overviews have been published that discuss recent technological developments of MS to enable analysis distinct proteoforms [42–44].

6.2 Antibody-based

Another high-throughput approach is the protein microarray (PMA), of which two different types exist: forward phase protein arrays (FPPA) and reverse phase protein arrays (RPPA). Given that antibodies can be raised to specifically recognize sequence variations or PTM, they enable measurement of selected proteoforms. In FPPA, protein antibodies are immobilized on an array in known positions, and samples are then printed on the array. If a particular proteoform is present in the sample, the proteoform binds to the antibody and after exposure to a secondary antibody, the abundance can be measured. Each slide is incubated with a single protein sample, but multiple proteins can be measured simultaneously depending on the number of antibodies printed on the slide.

The "reverse" version of the FPPA is the RPPA methodology. In RPPA, samples are first printed on the array, and subsequently each slide is stained with a single protein antibody, followed by a secondary antibody to amplify the signal. The downsides of RPPA are that all samples must be printed at the same time to avoid methodological barriers due to printing irregularities between batches, and that RPPA can only be used to detect proteins for which a strictly validated antibody is available. As there is no separation of the proteins according to molecular weight, it is crucial that antibodies are proven to be highly specific, selective and reproducible. Plus, RPPA is biased to proteins and isoforms for which a strictly validated antibody is available. On the other hand, RPPA requires only a small number of cells (approximately 3×10^5 cells to test 400 different antibodies), making it highly suitable for retrospective clinical applications. As it in addition analyzes all samples at once, it allows a direct comparison of protein abundance across samples.

7. Proteomics in acute leukemia

7.1 Disease-specific proteoform landscape of acute myeloid leukemia and acute lymphoid leukemia

Acute leukemia is a heterogeneous group of diseases both in terms of biology and prognosis. Classification into those arising from the myeloid or the lymphoid lineage is based on cytomorphology and cytochemistry, with further differentiation into specific subgroups based on morphology, immunophenotyping, cytogenetics, and molecular genetics of the acute leukemia cells. However, present classification systems are not adequate to differentiate between all subtypes and do not always accurately predict the clinical outcome. Whether changes in the leukemic cells that cause those differences are due to developmental, genetic, or environmental effects, they all are ultimately mediated by changes in protein abundance or modification. Therefore, we hypothesize that systematic comparative or differential proteomics can discover changes in the presence and quantity of individual proteoforms that underlie these cellular changes, and can add to current diagnostics, prognostics and therapeutics.

Assessment of the "diseased"-proteome compared to the proteome of the "normal/healthy" cells (e.g. CD34+, CD38+CD34+, CD38−CD34+; a discussion about the optimal normal comparator is discussed elsewhere [45]) can identify proteins that are aberrantly expressed or activated compared to normal, as well as can identify different forms of the same protein that differ between the diseased cell and the healthy comparator. This enables recognition of pathways utilization of cells present within a certain set of patients or related to a specific clinical feature. In addition, proteins or sets of proteins that are differentially expressed, may aid for confirmatory diagnostic purposes and early disease detection.

Furthermore, detailed proteomic profiling can help identifying differences between subgroups of diseases, including ALL and AML, and also between subgroups within one of both. It may be important (informative) to know how these two diseases are similar as well as how they differ. As ALL and AML are both dominated by immature malignant hematopoietic cells, they can serve as lineage-independent control for each other. Defining which proteins display similar expression in ALL and AML, but which are different compared to the "normal" healthy control, or to more mature cells, are likely to be related to a block in differentiation, whereas other proteins patterns that are similar in both, could be related to the hallmark of uncontrolled proliferation, resistance to cell death, or other shared deregulations.

As example, Cui et al. performed proteomic analysis using 2D-MS for 61 bone marrow biopsies from patients diagnosed with French-American-British (FAB) M1-M5 AML or ALL [46, 47]. Comparative analysis, identified 27 proteins with lineage-specific expression. Among them, myeloperoxidase was already known to be highly expressed in AML compared to ALL, but they also recognized heat shock factor binding protein 1 (HSBP1) as being high in ALL. In addition, they found proteins that were higher expressed in M2 and M3 AML compared to M1, and 23 proteins that were differentially expressed between granulocytic lineage (M1, M2, M3) AML, and AML derived from the monocytic lineage (M5). To prove clinical usefulness, Cui et al. also applied proteomic analysis to an AML-M3 bone marrow (which was classified based on morphology by the presence of atypical granules) from a patient who did not respond to the standard differentiation-inducing therapy with all-*trans* retinoic acid or As_2O_3. Their analysis showed that this sample exhibited a "protein expression profile" specific to M1, and not to M3, and after changing this treatment to chemotherapy, the patient gained complete remission within 3 weeks. Xu et al. performed proteomic profiling of the bone marrow samples from patients with different subtypes of acute leukemia (APL, AML, ALL) and healthy volunteers by SELDI-TOF-MS. Based on 109 protein signatures, they constructed a proteomic-based classification model capable of replicating the morphological and differentiation-based classification scheme of the well-established FAB system. Their results suggested that this mode could potentially serve as new diagnostic approach [48].

In our own group, we performed proteomic profiling using RPPA for 265 patients, in which we were able to separate 3 clusters of proteins that tended to track similarly within a FAB class from a subset of 24 differentially abundant proteins and PTMs; myeloid subtypes (M0–M2), the monocytic subtypes (M4–M5), erythroleukemia and megakaryocytic leukemia [49]. Foss et al. studied proteomics from 4 AML patients and 5 ALL patients using LC-MS/MS in blasts, as well as in CD34+ cells from 6 healthy donors and mononuclear cells from 2 healthy donors to correct for mononuclear cell contamination. Blinded unsupervised clustering enabled grouping with each cell type forming a discrete cluster, suggesting that proteomics can indeed, at least in some cases, robustly distinguish known classes of leukemia.

Recently, another study by our group analyzed pediatric AML (n = 95) and pediatric ALL (n = 73) on RPPA for antibodies against 149 different total proteins in addition to 45 antibodies recognizing different PTMs (e.g. phosphorylation, histone modification and cleavage) [50]. We felt that traditional hierarchical clustering was suboptimal as it weighs all proteins equally, in all situations across the dataset, and is agnostic to all known functional relationships between proteins, ignoring known interactions. Hence, we developed a novel computational method that accounts for known functional interactions which we call the "MetaGalaxy" approach [50–52]. This methodology starts with the allocation of proteins into groups of proteins with a related function based on existing knowledge or strong association within this dataset ("Protein Functional Group" (PFG), n = 31). For each PFG, a clustering algorithm enabled recognition of an optimal number of protein clusters; a subset of cases with similar (correlated) expression of core PFG components.

In order to know how the activity between the different PFG relate to each other within pediatric ALL and AML, we next hypothesized that there would be recurrent patterns of interaction between the various PFG clusters that would form a finite set of "protein expression signatures" that are shared by different subsets of patients. Therefore, patients were clustered based on their protein cluster membership using a binary matrix system. Correlation between protein clusters from various PFG was defined as a "Protein Constellation". We were able to identify subgroups of patients (signatures) that expressed similar combinations of protein constellations.

With this segmented approach a substantial amount of structure was observed across the data set (**Figure 1**), with an optimal number of 12 constellations and 12 signatures. Notably, signatures were strongly associated the leukemia-lineage. Signature 1 and 2 were specific to T-ALL (**Figure 1**, annotated in pink), whereas signatures 3, 4 and 5 were dominant to B-ALL and signature 7–12 to AML. Only signature 6 was a mixture of both B-ALL and AML patients. This clear distinction could also be discerned by the constellations. Protein constellation 1–4 were all specific to ALL, with constellation 1 (**Figure 1**, magenta box) only being found in B-ALL, 4 exclusively to T-ALL (**Figure 1**, yellow box), and 2 and 3 being present in both B- and T-ALL. On the other hand, constellation 7 and 8 were strongly associated with AML (**Figure 1**, blue box) and constellation 5 and 6 were found in both ALL and AML.

We also identified proteins that were universally changed in the same direction in at least 6 of the 8 signatures. Interestingly, GATA1 and STAT1 were universally

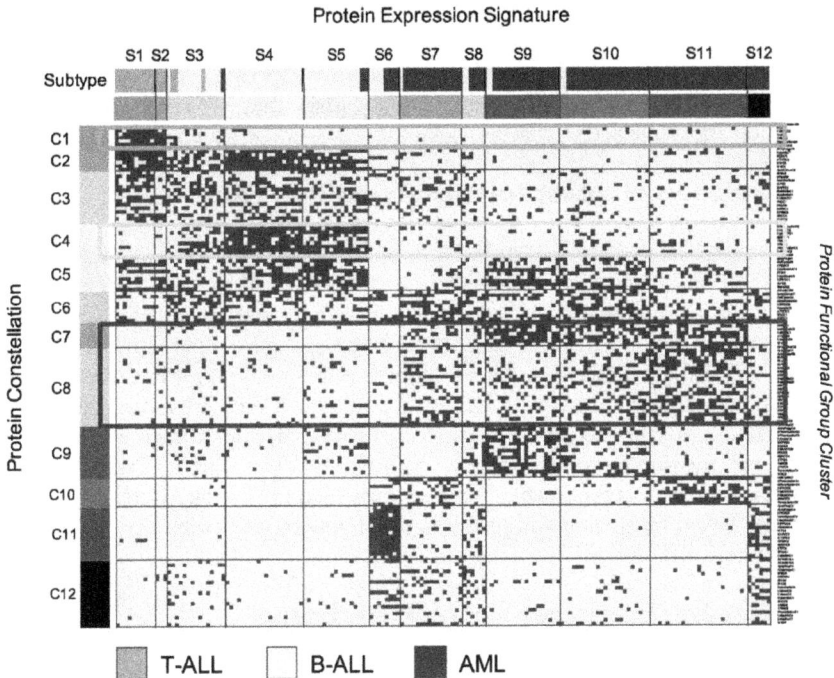

Figure 1.
"MetaGalaxy" analysis for pediatric ALL and AML. Annotations shows clear separation in protein patterns for T-ALL (magenta; signature 1 and 2), B-ALL (yellow; signature 3, 4, and 5), and AML (blue; signature 7, 8, 9, 10, 11, and 12). Constellations 1 (horizontally, magenta box) is associated with T-ALL, constellation 4 (horizontally, yellow box) with B-ALL and constellation 7 and 8 with AML (horizontally, blue box). This figure was adapted from Hoff et al. Molecular Cancer Research 2018 [50].

lower expressed in both pediatric AML and adult AML patients, whereas and phosphorylated RB1-pSer$^{807\text{--}811}$, a phosphorylation event that deactivates the RB1 protein, showed universally opposite expression in children and adults, being predominantly unphosphorylated (active) in pediatric patients and highly phosphorylated (inactive) in adults. For pediatric AML and ALL samples, comparable expressions were seen for the higher expressed universals CASP7 cleaved at domain 198 and phosphorylated CDKN1B-pSer^{10}, and the lower expressed JUN-pSer^{73} and GATA1.

Unpublished data from 205 adult AML and 166 adult ALL patients identified the existence of 11 protein signatures, of which 5 were AML dominant (93–100%), 4 were T-ALL dominant (79–100%) and 2 signatures contained a mixture of AML, B-ALL, T-ALL samples (50 and 68% AML). Three out of the 12 constellations were predominantly associated with AML, 4 were associated with ALL, 2 were associated with a mixture of ALL and AML cases, and 3 signatures were not strongly associated with any particular signature. This study used a total of 230 antibodies, including antibodies against 169 different proteins along with 52 antibodies targeting.

phosphorylation sites, 6 targeting Caspase and Parp cleavage forms and 3 targeting histone methylation sites. A third study (manuscript in preparation) from 500 pediatric AML, 68 adult T-ALL and 290 pediatric T-ALL patients, again showed similar results, with T-ALL and AML dominant signatures (81.5–100%), and only 1 out of the 15 signatures that had both T-ALL and AML (39% T-ALL and 61% AML). This clearly suggests that proteomics can be used to distinguish ALL from AML, and that although ALL and AML are very different in terms of overall proteomics, they share "protein expression signatures", which suggests that there are shared patterns of deregulation within some pathways. However, as all studies used a mixture of both total and PTM-proteins, it would be interesting to assess how expression and classification differs across diseases using a panel with a larger number of PTM, or a panel limited to PTM only, given that PTM often provide information about the activity or biological function of the protein.

7.2 Global proteomic landscape of pediatric and adult T-ALL

When we assessed the global proteomic landscape in pediatric and adult T-ALL, using the "MetaGalaxy" approach, we found 10 signatures based on 11 constellations (manuscript in preparation). Overall, signatures were not associated with age (i.e. pediatric vs. adult), with the exception of one signature. This signature was strongly associated with 2 constellations, which were only present in this particular signature. This suggests that pediatric T-ALL and adult T-ALL are more similar than ALL and AML, but that despite mostly overlapping signatures and constellations, there is an expression pattern specific to children. As this is similar to what we see in the genetics, were most recurrent aberrations are seen in both children and adults, but with different frequencies of occurring, correlation with genetic features would be interesting.

7.3 Assessing dynamic change upon treatment exposure

Children have a significant better prognosis and ALL responds better to treatment than AML. In addition to extracting information about differences in baseline protein abundance between those groups of patients, another consideration is to look at the dynamic response of the cells to stress, such as chemotherapy, or apoptotic inducers, to see whether changes in protein abundance patterns can provide a marker or whether a cell is responsive or resistant, and whether this is different between patients. Looking at post-treatment abundance and presence of proteoforms may provide insights into biological effects of drugs and mechanisms of drug resistance. This can either be done from static expression levels post-treatment at a given time point, or from the dynamic change in expression during treatment (i.e. expression post-treatment minus expression pre-treatment). Particularly, in leukemia, were blood can easily be drawn from the patient without performing any additional invasive procedures, expression can be measured at several time points during treatment.

Although this will not provide *a priori* information about which patients will respond to therapy or which patient needs which chemotherapy, it can give information about the response to treatment during early stages and so, aid in the decision of a more intensive treatment strategy should be achieved, or whether additional combinational treatment would be beneficial. For instance, if it is known that a particular protein pathway is utilized be the cell in order to circumvent cell death, in theory, this pathway can be targeted. Also, by comparing response to treatment on protein abundance or activity between ALL and AML, or children and adults, this can provide important information about why some patients respond while others do not.

While, theoretically, this approach would be promising, in reality this it much more complicated. First, of all, the time point of measuring the expression would be crucial. Assessment of the dynamic change too early, in cells that are not yet fatally hit by the chemotherapy or are in the process of dying, would suggest that the chemotherapy does not work, or has no effect on protein level, whereas measuring too late would measure the expression in cells that already died. Moreover, despite the ability of chemotherapy to kill the vast majority of leukemic cells, the rare leukemic stem cell that survives the chemotherapy, and that is responsible for the outgrowth of the leukemia cells which is manifested as relapse or primary resistant disease, is the cell from which we can potentially gather the most information. Proteomic analysis of these resistant cells, rather than taking the average of all, might be more informative than the analysis of the bulk leukemia population. Especially, knowing how those cells respond to chemotherapy (in comparison to other), would then be likely to raise new biological questions about why different cells behave differently, and why, or how, cells are able to circumvent chemotherapy, and what can be done to treat those cells. However, without a current means to *a priori* identify those few cells, isolation of (enough of) those cells remain a real challenge. So, if we want to know what is going in, pre- and post-treatment, as means to identify who those are, is required.

8. Conclusions

Despite significant improvement in treatment regimes, outcomes of both pediatric and adult patients with acute leukemia remain unsatisfactory. When a leukemia patient enters the clinic, particularly cytogenetics and mutation analysis are the methods of choice to perform risk stratification. And after induction therapy, choice of consolidation therapy is mainly based on the present chromosomal alternations and driver mutation(s). Emerging research in the field shows that prognosis is largely context-dependent and that acute leukemia are molecularly diverse diseases with similar phenotypes. Many years of exploration the molecular diversity in leukemia taught us that the combined influences of genetics, epigenetic remodeling, the microenvironment and PTM of leukemic blasts determine its cell fate. Since the net effect of these combined influences is predominantly displaced on the abundance and activity of the proteoforms, as well or their affected signaling pathways, we argue that characterization of differentially abundant proteoforms and recognition of proteomic patterns within and between (subgroups of) acute leukemia may facilitate and improve risk stratification as well as could provide therapeutic leads that may contribute to treatment personalization. However, while much is known about cytogenetics in AML and ALL, little is known about the proteomics of these cells.

While distinct proteoform patterns within and between different leukemic subtypes are only beginning to be recognized, age-specific proteome characterizations are far more limited. Bone marrow aspiration is a relative painful procedure and healthy donors, such as patient relatives or medical students who donate bone marrow that could function as internal control against AML blasts are scarce in many studies. The control group therefore often does not represent the median age of the patient cohort and leukemic-specific findings cannot be directly compared to a matched age group. Many studies focusing on leukemia therefore avoid controls and perform internal disease comparisons. Age-related analysis is then only applicable when a wide age distribution across the cohort is present, but this is often not the case as most research focuses on either pediatric or adult leukemia, instead of both.

More research is needed to identify single proteins and sets of proteins that are associated with disease and age specific subgroups. As far as we know, we are the first to analyze protein abundance and their PTM between AML and ALL across all ages, using antibody-based proteomics. Almost all studies look at AML or ALL and if they look at both, they mainly focus on the differences rather than the similarities. However, ALL and AML share the same pathophysiology in terms of the occurrence of a differentiation block that gives rise to uncontrolled clonal proliferations of immature hematopoietic progenitor cells in the bone marrow.

Defining which proteoforms have similar expression in ALL and AML, but different expression compared to the "normal" healthy control or to more mature cells are likely to be related to a block in differentiation. Other similar protein patterns could be related to the hallmark of an uncontrolled proliferation or resistance to apoptosis. Identification of differences in proteomic profiles between ALL and AML can additionally lead to lineage-specific proteomic signatures which may help to distinguish (subgroups) of the diseases.

Recognition of similar and dissimilar proteomic patterns among acute leukemia should also be analyzed in relation to responses to therapy. Treatment that is used in one group that was highly sensitive to it can be tested in other groups based on similar proteomic patterns. Cytogenetic and mutational information provides prognostic information, but so far lacks the *a priori* information to predict treatment outcomes. Rational selection of targeted therapies based on the functional activity state of the cell, as determined by the proteome, is more likely to sensitize patients for certain treatment regimens compared to random selection.

Conflict of interest

The authors declare no conflict of interest.

Author details

Fieke W. Hoff[1*†], Anneke D. van Dijk[1†] and Steven M. Kornblau[2*]

1 Department of Pediatric Oncology/Hematology, Beatrix Children's Hospital, University Medical Center Groningen, University of Groningen, Groningen, The Netherlands

2 Department of Leukemia, The University of Texas M.D. Anderson Cancer Center, Houston, TX, USA

*Address all correspondence to: f.w.hoff@umcg.nl and skornblau@mdanderson.org

† These authors contributed equally to the work.

IntechOpen

References

[1] Inaba H, Greaves M, Mullighan CG. Acute lymphoblastic leukaemia. The Lancet. 2013;**381**(9881):1943-1955

[2] Döhner H, Weisdorf DJ, Bloomfield CD. Acute myeloid leukemia. The New England Journal of Medicine. 2015;**373**(12):1136-1152

[3] Genshaft AS, Li S, Gallant CJ, Darmanis S, Prakadan SM, Ziegler CGK, et al. Multiplexed, targeted profiling of single-cell proteomes and transcriptomes in a single reaction. Genome Biology. 2016;**17**(1):188

[4] Payne SH. The utility of protein and mRNA correlation. Trends in Biochemical Sciences. 2015;**40**(1):1-3

[5] Vogel C, Marcotte EM. Insights into the regulation of protein abundance from proteomic and transcriptomic analyses. Nature Reviews. Genetics. 2012;**13**(4):227-232

[6] Katz AJ, Chia VM, Schoonen WM, Kelsh MA. Acute lymphoblastic leukemia: An assessment of international incidence, survival, and disease burden. Cancer Causes & Control: CCC. 2015;**26**(11):1627-1642

[7] Iacobucci I, Mullighan CG. Genetic basis of acute lymphoblastic leukemia. Journal of Clinical Oncology. 2017;**35**(9):975-983

[8] Boissel N, Baruchel A. Acute lymphoblastic leukemia in adolescent and young adults: Treat as adults or as children? Blood. 2018;**132**(4):351-361

[9] Van Vlierberghe P, Ferrando A. The molecular basis of T cell acute lymphoblastic leukemia. Journal of Clinical Investigation. 2012; **122**(10):3398

[10] Bhojwani D, Yang JJ, Pui CH. Biology of childhood acute lymphoblastic leukemia. Pediatric Clinics of North America. 2015;**62**(1): 47-60

[11] Pieters R, Schrappe M, De Lorenzo P, Hann I, De Rossi G, Felice M, et al. A treatment protocol for infants younger than 1 year with acute lymphoblastic leukaemia (Interfant-99): An observational study and a multicentre randomised trial. The Lancet. 2007;**370**(9583):240-250

[12] Pui C, Gaynon PS, Boyett JM, Chessells JM, Baruchel A, Kamps W, et al. Outcome of treatment in childhood acute lymphoblastic leukaemia with rearrangements of the 11q23 chromosomal region. The Lancet. 2002;**359**(9321):1909-1915

[13] Hunger SP, Loh ML, Whitlock JA, Winick NJ, Carroll WL, Devidas M, et al. Children's Oncology Group's 2013 blueprint for research: Acute lymphoblastic leukemia. Pediatric Blood & Cancer. 2013;**60**(6):957-963

[14] Rubnitz JE, Lensing S, Zhou Y, Sandlund JT, Razzouk BI, Ribeiro RC, et al. Death during induction therapy and first remission of acute leukemia in childhood: The St. Jude experience. Cancer. 2004;**101**(7):1677-1684

[15] Siegel RL, Miller KD, Jemal A. Cancer statistics, 2017. CA: A Cancer Journal for Clinicians. 2017;**67**(1):7-30

[16] Arber DA, Orazi A, Hasserjian R, Thiele J, Borowitz MJ, Le Beau MM, et al. The 2016 revision to the World Health Organization classification of myeloid neoplasms and acute leukemia. Blood. 2016;**127**(20):2391-2405

[17] Bullinger L, Döhner K, Döhner H. Genomics of acute myeloid leukemia diagnosis and pathways. Journal of Clinical Oncology. 2017;**35**(9):934-946

[18] Bolouri H, Farrar JE, Triche TJ, Ries RE, Lim EL, Alonzo TA, et al. The molecular landscape of pediatric acute myeloid leukemia reveals recurrent structural alterations and age-specific mutational interactions. Nature Medicine. 2018;**24**(1):103-112

[19] Ley TJ, Miller C, Ding L, Raphael BJ, Mungall AJ, Robertson AG, et al. Genomic and epigenomic landscapes of adult de novo acute myeloid leukemia. The New England Journal of Medicine. 2013;**368**(22):2059-2074

[20] Papaemmanuil E, Gerstung M, Bullinger L, Gaidzik VI, Paschka P, Roberts ND, et al. Genomic classification and prognosis in acute myeloid leukemia. The New England Journal of Medicine. 2016;**374**(23):2209-2221

[21] Gamis AS, Alonzo TA, Perentesis JP, Meshinchi S. Children's oncology Group's 2013 blueprint for research: Acute myeloid leukemia. Pediatric Blood & Cancer. 2013;**60**(6):964-971

[22] Gygi SP, Rochon Y, Franza BR, Aebersold R. Correlation between protein and mRNA abundance in yeast. Molecular and Cellular Biology. 1999;**19**(3):1720-1730

[23] Mun D, Bhin J, Kim J, Kim S, Kim HK, Kim DH, et al. Proteogenomic characterization of human early-onset gastric cancer. Cancer Cell. 2019;**35**(1):124.e10

[24] Zhang B, Wang J, Wang X, Zhu J, Liu Q, Shi Z, et al. Proteogenomic characterization of human colon and rectal cancer. Nature. 2014;**513**(7518):382-387

[25] Perez-Andreu V, Roberts KG, Xu H, Smith C, Zhang H, Yang W, et al. A genome-wide association study of

susceptibility to acute lymphoblastic leukemia in adolescents and young adults. Blood. 2015;**125**(4):680-686

[26] Zuurbier L, Petricoin EF, Vuerhard MJ, Calvert V, Kooi C, Buijs-Gladdines JGCAM, et al. The significance of PTEN and AKT aberrations in pediatric T-cell acute lymphoblastic leukemia. Haematologica. 2012;**97**(9):1405-1413

[27] Trinquand A, Tanguy-Schmidt A, Ben Abdelali R, Lambert J, Beldjord K, Lengliné E, et al. Toward a NOTCH1/FBXW7/RAS/PTEN–based oncogenetic risk classification of adult T-cell acute lymphoblastic leukemia: A group for research in adult acute lymphoblastic leukemia study. Journal of Clinical Oncology. 2013;**31**(34):4333-4342

[28] Bolouri H, Farrar JE, Triche T, Capone S, Ramsingh G, Ries RE, et al. The molecular landscape of pediatric acute myeloid leukemia reveals recurrent structural alterations and age-specific mutational interactions. Nature Medicine. 2019;**24**(1):103-112

[29] Winters AC, Bernt KM. MLL-rearranged leukemias—An update on science and clinical approaches. Frontiers in Pediatrics. 2017;**5**:1-21

[30] Liedtke M, Cleary ML. Therapeutic targeting of MLL. Blood. 2009;**113**(24):6061-6068

[31] Marcucci G, Mrózek K, Radmacher MD, Garzon R, Bloomfield CD. The prognostic and functional role of microRNAs in acute myeloid leukemia. Blood. 2011;**117**(4):1121-1129

[32] Noren Hooten N, Abdelmohsen K, Gorospe M, Ejiogu N, Zonderman AB, Evans MK. microRNA expression patterns reveal differential expression of target genes with age. PLoS One. 2010;**5**(5):e10724

[33] Daschkey S, Röttgers S, Giri A, Bradtke J, Teigler-Schlegel A, Meister G, et al. MicroRNAs distinguish cytogenetic subgroups in pediatric AML and contribute to complex regulatory networks in AML-relevant pathways. PLoS One. 2013;8(2):e56334

[34] Lawrence M, Daujat S, Schneider R. Lateral thinking: How histone modifications regulate gene expression. Trends in Genetics. 2016;32(1):42-56

[35] Ryan JM, Cristofalo VJ. Histone acetylation during aging of human cells in culture. Biochemical and Biophysical Research Communications. 1972;48(4):735-742

[36] Feser J, Tyler J. Chromatin structure as a mediator of aging. FEBS Letters. 2011;585(13):2041-2048

[37] Bhatla T, Wang J, Morrison DJ, Raetz EA, Burke MJ, Brown P, et al. Epigenetic reprogramming reverses the relapse-specific gene expression signature and restores chemosensitivity in childhood B-lymphoblastic leukemia. Blood. 2012;119(22):5201-5210

[38] Bacher U, Schnittger S, Macijewski K, Grossmann V, Kohlmann A, Alpermann T, et al. Multilineage dysplasia does not influence prognosis in CEBPA-mutated AML, supporting the WHO proposal to classify these patients as a unique entity. Blood. 2012;119(20):4719-4722

[39] Friedman AD. C/EBPα in normal and malignant myelopoiesis. International Journal of Hematology. 2015;101(4):330-341

[40] Aebersold R, Mann M. Mass spectrometry-based proteomics. Nature. 2003;422(6928):198-207

[41] Hillenkamp F, Karas M, Beavis RC, Chait BT. Matrix-assisted laser desorption/ionization mass spectrometry of biopolymers. Analytical Chemistry. 1991;63(24):1203A

[42] Cai W, Tucholski TM, Gregorich ZR, Ge Y. Top-down proteomics: Technology advancements and applications to heart diseases. Expert Review of Proteomics. 2016;13(8):717-730

[43] Lisitsa A, Moshkovskii S, Chernobrovkin A, Ponomarenko E, Archakov A. Profiling proteoforms: Promising follow-up of proteomics for biomarker discovery. Expert Review of Proteomics. 2014;11(1):121-129

[44] Schaffer LV, Millikin RJ, Miller RM, Anderson LC, Fellers RT, Ge Y, et al. Identification and quantification of proteoforms by mass spectrometry. Proteomics. 2019;19(10):e1800361

[45] Kornblau SM, Coombes KR. Use of reverse phase protein microarrays to study protein expression in leukemia: Technical and methodological lessons learned. Methods in Molecular Biology. 2011;785:141

[46] Cui J, Wang J, He K, Jin B, Wang H, Li W, et al. Two-dimensional electrophoresis protein profiling as an analytical tool for human acute leukemia classification. Electrophoresis. 2005;26(1):268-279

[47] Cui J, Wang J, He K, Jin B, Wang H, Li W, et al. Proteomic analysis of human acute leukemia cells: Insight into their classification. Clinical Cancer Research. 2004;10(20):6887-6896

[48] Xu Y, Zhuo J, Duan Y, Shi B, Chen X, Zhang X, et al. Construction of protein profile classification model and screening of proteomic signature of acute leukemia. International Journal of Clinical and Experimental Pathology. 2014;7(9):5569

[49] Kornblau SM, Tibes R, Qiu YH, Chen W, Kantarjian HM, Andreeff M,

et al. Functional proteomic profiling of AML predicts response and survival. Blood. 2009;**113**(1):154-164

[50] Hoff FW, CW H, Qiu Y, Ligeralde AA, Yoo SY, Mahmud M, et al. Recognition of recurrent protein expression patterns in pediatric acute myeloid leukemia identified new therapeutic targets. Molecular Cancer Research. 2018;**16**(8):1275-1286

[51] Hoff FW, Qiu Y, Kornblau SM, De Bont ESJM, Hu CW, Ligeralde A, et al. Recurrent patterns of protein expression signatures in pediatric acute lymphoblastic leukemia: Recognition and therapeutic guidance. Molecular Cancer Research. 2018;**16**(8):1263-1274

[52] Hu CW, Qiu Y, Ligeralde A, Raybon AY, Yoo SY, Coombes KR, et al. A quantitative analysis of heterogeneities and hallmarks in acute myelogenous leukaemia. Nature Biomedical Engineering. 2019;**3**(11):889-901

www.ingramcontent.com/pod-product-compliance
Lightning Source LLC
Chambersburg PA
CBHW081239190326
41458CB00016B/5837